Electromagnetics and Experimental Measurements of the Skin Effect

Other related titles:

You may also like

- PBPO203 | Zacharias | Inductive Devices in Power Electronics: Materials, measurement, design and applications | 2024
- SBEW550 | Araneo | Advanced Time Domain Modeling for Electrical Engineering | 2022

We also publish a wide range of books on the following topics:

Computing and Networks
Control, Robotics and Sensors
Electrical Regulations
Electromagnetics and Radar
Energy Engineering
Healthcare Technologies
History and Management of Technology
IET Codes and Guidance
Materials, Circuits and Devices
Model Forms
Nanomaterials and Nanotechnologies
Optics, Photonics and Lasers
Production, Design and Manufacturing
Security
Telecommunications
Transportation

All books are available in print via https://shop.theiet.org or as eBooks via our Digital Library https://digital-library.theiet.org.

IET Electromagnetic Waves 578

Electromagnetics and Experimental Measurements of the Skin Effect

Malcolm Stuart Raven

Institution of Engineering and Technology

British Library Cataloguing in Publication Data

A catalogue record for this product is available from the British Library

ISBN 978-1-83724-530-7 (hardback)
ISBN 978-1-83724-531-4 (PDF)

Typeset in India by MPS Limited

Cover image credit: Yuichiro Chino/Moment via Getty Images

Contents

Preface

Although there is a very large literature on theoretical electromagnetism, there is much less on experimental measurements, particularly for the *skin effect* at low frequencies. The author has found that of several recent papers he has written in this field, the one which received the most attention concerned experimental measurements of the skin effect at low frequencies [1,7]. For this reason, this monograph emphasizes the experimental measurements and results obtained for the skin effect, particularly at low frequencies. For a review of the subject, the article https://en.wikipedia.org/wiki/Skin_effect in Wikipedia is suggested. [2]. Presumably, in the case of AI, it must employ someone, perhaps like CERN, to conduct the experiments necessary to validate its conclusions.

However, to begin with, we need to establish the theoretical background necessary to understand the results obtained from measurements. The book therefore begins with an introductory Chapter 1 on the derivation of the equations of time-dependent electromagnetic disturbances, in which we compare James Clerk Maxwell's original approach [3] with the present-day approach. The remaining chapters mainly concern the solution of Maxwell's equations in the frequency domain.

The importance of Maxwell's discoveries has been widely recognized. The following quote from the introductory chapter may serve to encourage the reader. As Roger Penrose points out, "*Maxwell's equations* were the first of the relativistic equations" and "the theory of electromagnetism plays an important part in quantum theory, providing the archetypical field of *quantum field theory*" [4]. Also, Albert Einstein, stressed "The Special Theory of relativity has crystallized out from the Maxwell-Lorentz theory of electromagnetic phenomena", "which in no way opposes the theory of relativity" [5].

The book does not necessarily assume that the reader has a detailed knowledge of electromagnetism. It may therefore also be of interest to those involved in other disciplines where electromagnetism is not a major subject but includes fairly advanced mathematics at a level, say, of Arfken [6]. Other disciplines include geophysics, mechanical engineering, and mining engineering, where electromagnetic techniques are widely employed. Although there are other textbooks and papers directed towards this goal, the approach employed here follows Maxwell's original analysis [3], which is not usually the case. This approach also does not require a detailed knowledge of electrical engineering, but it does lead to the fundamental equations of electromagnetism; the diffusion equations in conductors and the wave equations in non-conductors, leading to the electromagnetic theory of light and the pressure exerted by electromagnetic radiation.

Chapter 2 introduces Maxwell's four vector equations (all based on experimental measurements) with solutions, initially for free space, leading to the wave equations for electric and magnetic fields, travelling waves, and the relationship between the electric and magnetic fields, plane waves with two components, and the constants of propagation for free space. This is followed by a section on the solutions of Maxwell's equations for lossy materials. This again leads to wave equations but with additional diffusion terms - the Helmholtz Wave Equations or Equations of Telegraphy, details of the propagation constants in lossy materials, complex refractive index, optical constants, Debye equations, dissipation factor, and circuit parameters.

Chapter 3 concerns steady state and time-dependent power dissipation, including power dissipation in circuits, power dissipation in the time and frequency domain, power factor, instantaneous power, oscillatory power (important in inductive circuits), power flow, poynting theorem, superconductivity, complex poynting theorem, relaxation effect, impedance, dissipation in circuits. This chapter and Chapter 2 also briefly discuss superconductors.

Chapter 4 is an Introductory chapter to the skin effect, including approximate methods of analysis for various conductor geometries. It begins with a general explanation about the decrease of the electromagnetic fields with depth as the frequency increases, the use of high conductivity films on conductors, hollow conductors at low frequencies, and high permeability materials. This chapter also includes a brief history of the skin effect, which was very important to the pioneers of radio communications. In recent times, the large development of wind turbines and the extension of the electrical transmission lines to the many remote sites have led to increased energy losses due to the low-frequency skin effect.

This is followed by chapters covering more detailed theory and experimental measurements of the skin effect in solid and hollow cylindrical tube conductors. Copper or aluminium tube busbars are used in electricity substations and have many advantages over solid copper busbars. These busbars must withstand very high current and voltage switching transients. The final chapter describes methods of measuring the skin effect over a wide range of frequencies. In addition to the dedicated techniques used to measure L, C, and R, a Gain Phase-Meter (GPM) technique was also employed here to measure the amplitude and phase to determine the impedance as a function of frequency. This was particularly useful for the low-frequency skin effect, where the resistance may be less than a milliohm.

The book finishes with an Appendix containing Bessel's modified equation, Kelvin functions, properties of Bessel functions, power integral and orthogonality, and finally, a reference section and index.

Copper tube theoretical impedance and internal inductance.

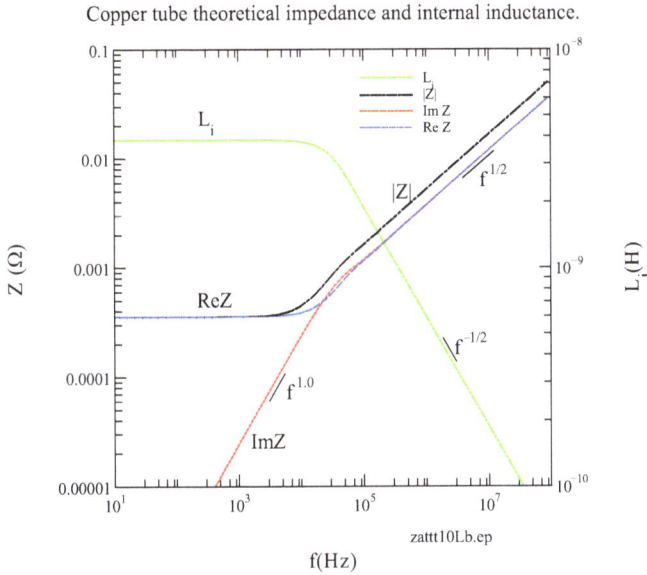

Figure P.1   *Frequency dependance of theoretical impedance and inductance for a hollow copper tube showing skin effect. In this example the tube length was 3.04 m with inner radius 4.3 mm and outer radius 5 mm. See Chapter 11, (11.1) for further details.*

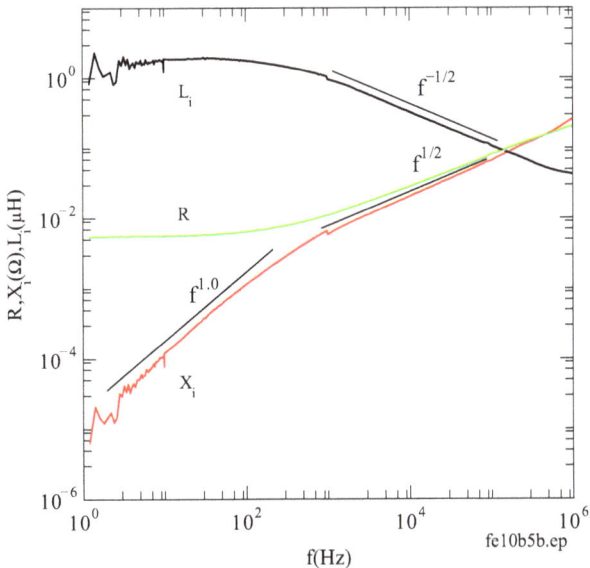

Figure P.2   *Iron wire, diameter 2 mm, formed into a rectangular loop with mean gap width 18 mm, length 19.2 cm and current $I = 0.6$ Arms. Measurements performed using a Gain Phase-Meter (GPM) method [36] with function generator HP3325A and low- frequency power amplifier for frequency range 10 Hz to 1 kHz and high-frequency power amplifier for 1 kHz to 1 MHz. Internal inductance (L_i), resistance (R) and reactance (X_i). See pp. 101 and 117.*

About the author

Malcolm Stuart Raven is engaged in theoretical and experimental work on aspects of the skin effect in solid and tubular conductors, and the development of electronic measuring techniques, operating from his own laboratory in the UK. From 1978 to his retirement in 2001, he was a lecturer in the department of Electrical and Electronic Engineering at the University of Nottingham. His research and publications mainly concerned the growth and analysis of thin films and devices and high-temperature superconducting thin films. This included teaching in the USA and presenting conference and seminar talks to universities and industry in the UK, Europe and Japan. He also worked in collaboration with the physics department at Nottingham and was a founding member of the Nottingham University Molecular Beam Epitaxy Research Syndicate (NUMBERS).

Chapter 1

Maxwell's general equations of electromagnetic disturbances

The theory of the propagation of electromagnetic disturbances in the time domain provides the fundamental explanation of how electrical energy is transmitted both for the steady state dc case and steady state sinusoidal transmission. This is not only applicable to modern communication systems in electronic engineering and physics but also in other disciplines including geophysics and mechanical engineering. In this chapter we firstly review the basic theory based on Maxwell's equations and then apply it to examples including diffusion in conductors and ferromagnetic conductors.

1.1   Introduction

This chapter concerns the derivation and equations of time dependent electromagnetic disturbances in which we compare James Clerk Maxwell's original approach with the present day approach. In the following chapters we consider Maxwell's equations reduced to the well known four vector calculus equations first determined by Oliver Heaviside in 1884 and recently reviewed by D.P. Hampshire [8].

However, the chapter does not necessarily assume that the reader has a detailed knowledge of electromagnetism. It serves to introduce important topics and equations in electromagnetism. It may therefore be of interest to those involved in other disciplines where electromagnetism is not their major subject. This may include disciplines, such as geophysics, mechanical and mining engineering. Although there are other textbooks and papers directed towards this goal, the approach employed here follows Maxwell's original analysis [3,17,21], which is not usually the case. Physicists and electrical engineers may also find the vector potential method employed by Maxwell of interest. Using this approach in his theory of light, he revealed the fundamental wave equations and diffusion equations of electromagnetism.

The importance of Maxwell's discoveries has been widely recognized. As Roger Penrose points out, '*Maxwell's equations* were the first of the relativistic equations' and 'the theory of electromagnetism plays an important part in quantum theory, providing the archetypal field of *quantum field theory*', [4]. Also, as Albert Einstein stressed, 'The Special Theory of relativity has crystallized out from the Maxwell–Lorentz theory of electromagnetic phenomena', 'which in no way opposes the theory of relativity.' [5].

We initially review Maxwell's Chapter 20 in Maxwell's *Treatise on Electricity and Magnetism* [3,17,21]. Although this chapter is entitled *Electromagnetic Theory of Light*, Maxwell first derived the general equations for an electromagnetic disturbance through any uniform medium at rest. For the case of non-conductors, Maxwell produced the wave equations, calculated the speed of light, which led to the understanding that light arises from electromagnetic waves. In addition, he calculated the pressure and momentum of light, which led to the present notion of photons and wave mechanics. For conductors, the solutions lead to diffusion equations, the penetration of electromagnetic waves, and the important topic of the *Skin Effect*. These have applications in many areas of science and engineering.

This is followed by the alternative analysis, which uses the Lorenz gauge to derive the wave equations and diffusion equation with sources. This leads to the calculation of the impedance of a conductor in the time domain in terms of vector potentials.

1.2   Basics of the transient transmission of electrical energy

Lightning is a large electrical transient that involves the rapid discharge of clouds to the earth (ground) or between clouds. The cloud (Cumulonimbus) is initially charged by a complex process of molecular friction. Eventually, the cloud's electrical potential to ground reaches the breakdown strength of the surrounding air (about 30 kV/cm), and the cloud discharges its electricity to ground with a bang (thunder). Although the details of the complete process are very complex, involving most of the known Physics and maybe more, lightning provides a dramatic example of transient electrical phenomena [9].

Initially, it is worth considering a simpler case of how an electrical disturbance travels along conducting wires when power is switched on to a load such as a light bulb. The bulb appears to light immediately when the switch is turned on, even though it may be some distance from the power source. But the drift velocity of electrons for 1 amp through, say, a section 1mm × 1mm of copper wire is only about 26 cm/hr or 2.6 m per 10 hours – the pace of a very tired snail. It would take a day or more for the light to come on! Hence this cannot be the only explanation since we know that the current can easily be measured flowing through the bulb or anywhere along the wire the instant that the light is switched on. One explanation of how the current can travel much faster than the drift current is that the electrons in the conductor only move about a mean position as the wave passes, rather like the 'Mexican Wave' at a football match where fans raise their arms in succession and a wave of arms travels around the stadium [10,11].

The full theoretical explanation turns out to be quite complex, involving the coupling of the electric and magnetic fields in the conductor and in the surrounding space. The EM field and displacement current outside the wire travel at nearly the speed of light, leading to radiation of the *E* and *H* fields. This is commonly detected by radios when the light is switched on or lightning strikes.

In the theory based on Maxwell's equations, only macroscopic effects are considered, where the electric and magnetic fields are averaged over a microscopic region [12] or averaged over quantum properties. In this case, the nearly instant effect is explained as due to an electromagnetic wave producing the displacement current in the insulation outside the conductors, and the wave travels at nearly the speed of light to the load driving current through the bulb. The current that penetrates the conductors, however, obeys a diffusion equation. If the wave is sinusoidal or pulsed, the current may not have had time to penetrate to the centre of the conductor before the wave reverses or the pulse falls. This leads to the notion of the *Skin Effect*, which concerns the finite depth of penetration of time-dependent currents into a conductor. For copper at 50 Hz, the *calculated* skin depth $\delta = 9.28$ mm and the phase velocity $u = 2.9$ m/s. At 1 MHz $\delta = 65.6$ μm and $u = 412$ m/s. Although this is much less than the speed of light, the current only has to diffuse to half the diameter of the conductor rather than the full length of the cable [13].

1.2.1   *Electromagnetic disturbances-frequency and time domains*

The topic of electromagnetic disturbances is mostly considered from two general perspectives: disturbances as a function of frequency (frequency domain electromagnetics) and disturbances as a function of time (time domain electromagnetics). A large proportion of the literature on this subject considers the frequency domain. This is mainly because the history of the subject is dominated by electromagnetic communications in the frequency domain. This is also convenient for the mathematical analysis since it leads to simpler equations with analytical solutions. However, the theory of disturbances in the time domain provides the fundamental explanation of how electrical energy is transmitted both for the steady-state dc case and steady-state sinusoidal transmission.

In addition, disturbances in the time domain are of great technological importance because of the rapid development in digital electronics and digital communications. This aspect of transient electromagnetics includes natural phenomena such as lightning and applications in geophysics to gain information on the electrical resistivity of subsurfaces [14]. There are also many common examples of transient behaviour such as that which occurs when switching lights on or off, and in dc systems such as switching on vehicles, transient arcing in railway systems [15] or any electrical appliance.

Generally, the theory is approached in two ways: (1) EM field theory that solves Maxwell's equations under particular boundary conditions. (2) Transmission line theory that considers the conductors as elements of inductance, capacitance and resistance. The first method essentially considers the physics of the problem and is more suitable for complex boundary conditions and numerical analysis. The second method has the advantage that it can be taught as an extension of circuit theory and is widely used in electrical engineering courses. In the case of circuits with only linear inductance, capacitance and resistance, a second-order linear differential equation can be established and the circuit solved analytically using an integrating factor or by using Laplace Transforms. If it is necessary to take into account frequency dependence of the inductance, capacitance, resistance or skin effect in the conductors, then the problem becomes more challenging, and numerical methods are frequently used [108,109].

1.3   Theory of the propagation of electromagnetic disturbances – Maxwell's approach

In Maxwell's analysis vector symbols are represented by Euler Fraktur fonts [16] (see Appendix section 1.10) – mainly because as he said 'the number of different vectors being so great that Hamilton's favourite symbols would have been exhausted at once' (Art. 618).* Although these look rather elegant compared to present-day fonts, they make appreciating the equations rather difficult. I have therefore presented the equivalent equations in SI form alongside Maxwell's equations.

As mentioned, the theory assumes that the electric and magnetic fields are averaged over a microscopic region of the medium [12]. The total current flowing through the medium at rest due to an electromagnetic disturbance is given by the sum of the conduction current and the displacement current.

$$\text{Maxwell: } \mathfrak{C} = \left(C + \frac{1}{4\pi} K \frac{\partial}{\partial t} \right) \mathfrak{E}, \quad \text{SI: } \mathbf{J} = \left(\sigma + \epsilon \frac{\partial}{\partial t} \right) \mathbf{E} \tag{1.1}$$

total current density $\mathfrak{C}$ and $\mathbf{J}$, specific conductivity C and σ, inductive capacity K and permittivity $\epsilon = \epsilon_o \epsilon_r$, electric field $\mathfrak{E}$ and $\mathbf{E}$. The notion of an additional current due to the time dependent fields, the *displacement current* was a significant contribution which allowed Maxwell to develop his theory of light, Maxwell (Art. 610), [18,19].

If the conductor is moving we need to add a third electric field due to the conductor moving through the magnetic field which is $E_m = u \times B$.

$$\mathfrak{E} = \mathfrak{G} \times \mathbf{B} - \nabla \Psi - \frac{\partial \mathfrak{U}}{\partial t}, \quad \mathbf{E} = \mathbf{u} \times \mathbf{B} - \nabla \Psi - \frac{\partial \mathbf{A}}{\partial t} \tag{1.2}$$

where $\mathbf{u} = \mathfrak{G}$ is the velocity of the conductor Ψ is a scalar electric potential and $\mathfrak{U}$ and $\mathbf{A}$ are vector potentials defined by $\mathbf{B} = \nabla \times \mathbf{A}$. Thus, the total electric field in the time domain is the sum of the motional electric field, the gradient of the electric potential and the time variation of the magnetic vector potential, [3] Art. 599.

If there is no motion in the medium the electromotive intensity is

$$\mathfrak{E} = -\nabla \Psi - \frac{\partial \mathfrak{U}}{\partial t}, \quad \mathbf{E} = -\nabla \Psi - \frac{\partial \mathbf{A}}{\partial t} \tag{1.3}$$

This equation for the electric field is very important, and a proof is given in Section 1.9.1. Substituting this into (1.1) gives

$$\mathfrak{C} = -\left(C + \frac{1}{4\pi} K \frac{\partial}{\partial t} \right) \left(\nabla \Psi + \frac{\partial \mathfrak{U}}{\partial t} \right), \quad \mathbf{J} = -\left(\sigma + \epsilon \frac{\partial}{\partial t} \right) \left(\nabla \Psi + \frac{\partial \mathbf{A}}{\partial t} \right) \tag{1.4}$$

But Maxwell uses a relation for the current density given in Art. 616

$$4\pi \mu_1 \mathfrak{C} = \nabla^2 \mathfrak{U} + \nabla J_M, \quad \mu \mathbf{J} = -\nabla^2 \mathbf{A} + \nabla \nabla.(\mathbf{A}) \tag{1.5}$$

where $\mu_1 = \mu/(4\pi)$ and $\mu = \mu_o \mu_r$. The proof of (1.5) is given in Section 1.9.3. Note that Maxwell used a negative Laplacian $(-\nabla^2)$ in determining (1.5), which gives rise

---

*Numbered articles in reference are referred to by Art.nnn.

to a positive value for $\nabla^2 \mathfrak{U}$. We have used the conventional positive value which gives a negative value for $\nabla^2 \mathbf{A}$ in (1.5). Maxwell defined J_M as

$$J_M = \frac{\partial F}{\partial x} + \frac{\partial G}{\partial y} + \frac{\partial H}{\partial z} \tag{1.6}$$

$$\text{or } J_M = \frac{\partial A_x}{\partial x} + \frac{\partial A_y}{\partial y} + \frac{\partial A_z}{\partial z} = \nabla . \mathbf{A}$$

where F, G, H and A_x, A_y, A_z are the vector potential components in the x, y and z directions, respectively. In this equation we have given Maxwell's 'J' a suffix M to distinguish it from the SI unit for current density $\mathbf{J}$.

Combining (1.4) and (1.5) gives

$$\mu_1 \left(4\pi C + K\frac{\partial}{\partial t} \right) \left(\frac{\partial \mathfrak{U}}{\partial t} + \nabla \Psi \right) + \nabla^2 \mathfrak{U} + \nabla J_M = 0, \tag{1.7}$$

$$\mu \left(\sigma + \epsilon \frac{\partial}{\partial t} \right) \left(\frac{d\mathbf{A}}{\partial t} + \nabla \Psi \right) - \nabla^2 \mathbf{A} + \nabla (\nabla . \mathbf{A}) = 0$$

Expanding this equation yields three equations

$$\mu_1 \left(4\pi C + K\frac{\partial}{\partial t} \right) \left(\frac{\partial F}{\partial t} + \frac{\partial \Psi}{\partial x} \right) + \nabla^2 F + \frac{\partial J_M}{\partial x} = 0$$

$$\mu_1 \left(4\pi C + K\frac{\partial}{\partial t} \right) \left(\frac{\partial G}{\partial t} + \frac{\partial \Psi}{\partial y} \right) + \nabla^2 G + \frac{\partial J_M}{\partial y} = 0 \tag{1.8}$$

$$\mu_1 \left(4\pi C + K\frac{\partial}{\partial t} \right) \left(\frac{\partial H}{\partial t} + \frac{\partial \Psi}{\partial z} \right) + \nabla^2 H + \frac{\partial J_M}{\partial z} = 0$$

These are Maxwell's '*general equations of electromagnetic disturbances*', which he then differentiated with respect to x, y and z and summed to obtain, after some work,

$$\mu_1 \left(4\pi C + K\frac{\partial}{\partial t} \right) \left(\frac{\partial J_M}{\partial t} - \nabla^2 \Psi \right) = 0 \tag{1.9}$$

$$\mu \left(\sigma + \epsilon \frac{\partial}{\partial t} \right) \left(\frac{\partial (\nabla . \mathbf{A})}{\partial t} + \nabla^2 \Psi \right) = 0$$

Then

$$\left(\sigma + \epsilon \frac{\partial}{\partial t} \right) \nabla . \left[\frac{\partial \mathbf{A}}{\partial t} + \nabla \Psi \right] = 0 \tag{1.10}$$

Substituting E from (1.3) gives

$$\left(\sigma + \epsilon \frac{\partial}{\partial t} \right) \nabla . (-\mathbf{E}) = 0 \tag{1.11}$$

Using Maxwell's first equation

$$\left(\sigma + \epsilon \frac{\partial}{\partial t} \right) (-\rho_v / \epsilon) = 0 \tag{1.12}$$

Hence

$$\frac{\partial \rho_v}{\partial t} + \frac{\sigma \rho_v}{\epsilon} = 0 \tag{1.13}$$

which is the continuity equation for an open surface.

1.3.1   Conductors

If the medium is a conductor with very small displacement current compared with the conduction current then Maxwell in Art. 783 put $\nabla \Psi = \nabla J_M = 0$ so that (1.8) becomes

$$\mu_1 4 \pi C \frac{\partial \mathfrak{U}}{\partial t} + \nabla^2 \mathfrak{U} = 0, \tag{1.14}$$

$$\nabla^2 \mathbf{A} - \mu \sigma \frac{d\mathbf{A}}{\partial t} = 0$$

These are source free homogeneous diffusion equations with known solutions [3,20,23].

1.3.2   Non-conductors

Maxwell argued that for '*a non-conductor, $C = 0$ and $\nabla^2 \Psi$ which is proportional to the volume density of free electricity, is independent of t. Hence J_M must be a linear function of t or a constant or zero*'. Hence for periodic disturbances J_M and Ψ were set to zero. Equation (1.8), then becomes

$$\nabla^2 F + \mu_1 K \frac{\partial^2 F}{\partial t^2} = 0$$

$$\nabla^2 G + \mu_1 K \frac{\partial^2 G}{\partial t^2} = 0 \tag{1.15}$$

$$\nabla^2 H + \mu_1 K \frac{\partial^2 H}{\partial t^2} = 0$$

The condition in which J_M or $\nabla . \mathbf{A} = 0$ is referred to as the *Coulomb gauge* [22]. An alternative analysis that uses the Lorenz gauge is given in a following section.

Equation (1.10) is a source-free homogeneous wave equation for the vector potential components F, G and H in the x, y and z directions, respectively. The positive sign arises because Maxwell used $-\nabla^2$. Maxwell then proceeded to show that the velocity of the waves was close to the measured velocity of light suggesting that light was an electromagnetic disturbance. Equation (1.16) gives the same equations in SI form where A_x, A_y and A_z are the components of the vector potential $\mathbf{A}$

$$\nabla^2 A_x - \mu \epsilon \frac{\partial^2 A_x}{\partial t^2} = 0$$

$$\nabla^2 A_y - \mu \epsilon \frac{\partial^2 A_y}{\partial t^2} = 0 \tag{1.16}$$

$$\nabla^2 A_z - \mu \epsilon \frac{\partial^2 A_z}{\partial t^2} = 0$$

1.4 Electromagnetic waves and light

The equation which represents a periodic wave travelling in the +z-direction is given by

$$\frac{\partial^2 f}{\partial z^2} = (1/u^2)\frac{\partial^2 f}{\partial t^2} \tag{1.17}$$

where f is the amplitude and u the velocity of a fixed point on the wave, that is, the phase velocity. Comparing (1.17) with (1.16), then these equations represent periodic waves travelling at a phase velocity

$$u_o = \sqrt{\frac{1}{\mu_o \epsilon_o}} \tag{1.18}$$

In terms of the classical dielectric and magnetic constants where $\mu_1 = 1$ then $v = 1/\sqrt{K}$ or using electrostatic units $K = 1$, $v = 1/\sqrt{\mu}$, Maxwell found that u_o was close to the value c of the velocity of light measured at that time[†]. Subsequent measurements have confirmed the theory that light consists of electromagnetic waves. This theory was first given by Maxwell in 1865 [24]. Presently, the speed of light in vacuo is defined in SI units as $c = 2.99\ 792\ 458 \times 10^8$ ms^{-1} [25].

Based on the classical electrical measurements Art. 787 and [26], (1.18), the average value of 12 calculations of the velocity is $\bar{v}_c = 2.96315$ in units of 10^8 ms^{-1}. The average value of 6 directly measured light velocities was $\bar{v}_m = 2.998238$. These figures are summarized in Table 1.1. The results of Rosa and Newcombe are the individual values closest to the present-day velocity of light.

Table 1.1   *Average and individual determinations of the velocity of light from classical electrical measurements and classical direct measurements (in units of 10^8 ms^{-1}), Art. 787 [26]. Figures in brackets are percentage differences relative to the present value of 2.997924 [25].*

Electrical $\bar{v}_c$	Direct $\bar{v}_m$
2.96315	2.998238
(−1.16)	(0.01046)
Rosa	Newcombe
2.9993	2.99766
(0.0488)	(−0.000827)

[†]Conventionally c is referred to as the *speed* of light because it is constant and a scalar. Maxwell referred to it as the *velocity* of light.

1.5   Energy in electromagnetic waves – radiation pressure

In his theory of light, Maxwell also showed that the energy in electromagnetic radiation produced a stress, which led to his calculation of the pressure of sunlight [3] Art. 792. J.H. Poynting investigated this in detail a few years later. Although Poynting is well known for his theory of power flow in electromagnetic fields and the *Poynting Vector* [57], he also published a number of papers on the pressure due to electromagnetic radiation. The first of these was published in 1903 [29]. A comprehensive review of these papers has been given recently [30].

The pressure of electromagnetic radiation continues to be of significant interest because it implies that electromagnetic waves have momentum. This supports quantum theory, which explains that radiation has particle-like properties – photons. Radiation pressure is also of interest in astronomy since it helps explain why the tail of comets is deflected away from the sun. Also, in the design of satellites and other spacecraft, the effect of the sun's radiation pressure has to be taken into account.

1.5.1   Maxwell on the pressure of sunlight

Maxwell in Art. 792 gives 'the electrostatic energy per unit volume at any point of the wave in a non-conducting medium as'

$$w_E = \frac{K}{8\pi} \left(\frac{dF}{dt}\right)^2 \; ft.lb.ft^{-3}, \quad <w_E> = \frac{\epsilon E_{x1}^2}{4} \; Jm^{-3} \tag{1.19}$$

$$w_M = \frac{1}{8\pi\mu} \left(\frac{dF}{dt}\right)^2 \; ft.lb.ft^{-3}, \quad <w_M> = \frac{\mu H_{y1}^2}{4} \; Jm^{-3} \tag{1.20}$$

where the first equations, in (1.19) and (1.20) are from Maxwell, Art. 792 given in British units and the second equations are time averaged energy densities in SI units. F is the vector potential (A_x), E_{x1} and H_{y1} are the wave amplitudes. For plane waves in free space Maxwell gives (Art. 790)

$$\frac{d^2F}{dz^2} = K\mu\frac{d^2F}{dt^2} \tag{1.21}$$

Hence the ratio $w_E/w_M = 1$. For the SI case and free space with characteristic impedance Z_o the ratio of these energies is

$$\frac{<w_E>}{<w_M>} = \frac{\epsilon}{\mu} \frac{E_{x1}^2}{H_{y1}^2} = Z_o^2/Z_o^2 = 1 \tag{1.22}$$

The two energies are therefore 'equal for a single wave, that is, at every point on the wave the intrinsic energy of the medium is half electrostatic and half electrokinetic'. Maxwell then obtained the pressure in the waves by considering the tensions and pressure in the electrostatic and electrokinetic fields which he had earlier analyzed in great detail including Art. 107 and Art. 643. The following quotation from Art. 792 essentially provides in words a summary of this analysis.

> Let p be the value of either w_E or w_M, then for the electrostatic state of the medium, there is a tension p in direction parallel to x, combined with

a pressure p in direction parallel to x and z. For the electrokinetic state of the medium, there is a tension p in direction parallel to y, combined with a pressure p in direction parallel to x and z. The combined effect of the electrostatic and electrokinetic stresses is a pressure equal to $2p$ in the direction of propagation of the waves. Now $2p$ also expresses the whole energy in units of volume. Hence, in a medium in which waves are propagated, there is a pressure in the direction normal to the waves and numerically equal to the energy in units of volume.

For strong sunlight falling on one square foot, Maxwell used a value of 83.4 ft.lb.s^{-1}ft^{-2} or 1.217 kWm^{-2}. This value is close to 1.228 kWm^{-2} as measured by Pouillet [31].

The work done per second, that is, energy dissipated per second is

$$dW/dt = Fdr/dt = Fv \tag{1.23}$$

where the force F or rate of change of momentum is in the same direction as the velocity v. The pressure is then

$$p = \frac{F}{A} = \frac{d(mv)}{Adt} = \frac{1}{Av}\frac{dW}{dt} \tag{1.24}$$

Maxwell determined the velocity of light from the ratio of electric units (Art's 784, 787 $v = 1/\sqrt{K\mu}$) giving $v = 2.88 \times 10^8$ ms^{-1} or 9.446×10^8 ft.s^{-1}. Using his value of sunlight energy above gives the average pressure $p = 83.4/(9.446 \times 10^8) = 8.83 \times 10^{-8}$ lb.f. Converting this to SI units gives $p = 4.227 \times 10^{-6}$ Nm^{-2}. The maximum pressure is $2p = 8.45 \times 10^{-6}$ Nm^{-2}.

This compares with the present average pressure calculated using a value for the solar radiation (*Solar Constant*) of value 1.361 kWm^{-2} [32,33], which gives $p = 1361/(3 \times 10^8) = 4.537 \times 10^{-6}$ Nm^{-2} or maximum pressure $2p = 9.073 \times 10^{-6}$ Nm^{-2}. These results are summarized in Table 1.2. For convenience Table 1.3 lists the units used by Maxwell with equivalent SI units.

1.5.2   Radiation pressure and wave-particle duality

The pressure of electromagnetic radiation continues to be of significant interest because it implies that electromagnetic waves have momentum. This supports quantum theory which explains that radiation has particle-like properties – photons. In a simplified view of the duality between particles and waves, the pressure or

Table 1.2   *Sunlight radiation pressure: Comparison of Maxwell's calculation [3,17,21] with present day results [33]*

Ref.	Incident energy kWm^{-2}	Speed of light c 10^8 ms^{-1}	Maximum pressure 10^{-6} Nm^{-2}
Maxwell	1.217	2.88	8.45
Present	1.361	3	9.07

Table 1.3   British (fps) units used by Maxwell compared
with SI and cgs units

	British (fps)	**SI**	**cgs**
Force	1 lbf	4.448 N	4.448×10^5 dynes
Pressure	1 lbf ft.$^{-2}$	47.88 Nm^{-2}	478.8 dynes cm^{-2}

force applied to a particle mass m is, by Newton's first law, the rate of change of the particle's momentum (1.24). Thus, electromagnetic radiation may be described in terms of energy per second (wave) or change in momentum (particle) [4,99]. Recently, experiments have been carried out which demonstrate the equivalence of wave-particle duality with entropic uncertainty relations of quantum mechanics [113].

The subject of quantum optics has recently been reviewed by Barnett [114]. This article is part of a theme *The quantum theory of light* published in the same journal.

1.6   Alternative derivation – the Lorentz condition (gauge)

As an alternative approach relating the time dependent current density $\mathbf{J}(t)$ to the time dependent magnetic field $\mathbf{H}(t)$ from Ampere's law or Maxwell's equation

$$\mathbf{J}(t) = \nabla \times \mathbf{H}(t) = (1/\mu)\nabla \times \mathbf{B}(t) \tag{1.25}$$

$$= (1/\mu)\nabla \times \nabla \times \mathbf{A} = (1/\mu)[\nabla(\nabla.\mathbf{A}) - \nabla^2\mathbf{A}] \tag{1.26}$$

The time dependent current density which includes the displacement current is given by (1.4)

$$\mathbf{J}(t) = -\left(\sigma + \epsilon\frac{\partial}{\partial t}\right)\left(\frac{\partial \mathbf{A}}{\partial t} + \nabla\Psi\right) \tag{1.27}$$

which expands to

$$\mathbf{J}(t) = -\left(\sigma\nabla\Psi + \epsilon\frac{\partial\nabla\Psi}{\partial t} + \sigma\frac{\partial \mathbf{A}}{\partial t} + \epsilon\frac{\partial^2 \mathbf{A}}{\partial t^2}\right) \tag{1.28}$$

Equating with (1.26) gives

$$-\left(\sigma\nabla\Psi + \epsilon\frac{\partial\nabla\Psi}{\partial t} + \sigma\frac{\partial \mathbf{A}}{\partial t} + \epsilon\frac{\partial^2 \mathbf{A}}{\partial t^2}\right) = (1/\mu)[\nabla(\nabla.\mathbf{A} - \nabla^2\mathbf{A}] \tag{1.29}$$

This is the same as (1.7) after expanding the terms in brackets. Re-writing (1.29)

$$-\sigma\nabla\Psi - \epsilon\frac{\partial\nabla\Psi}{\partial t} = \sigma\frac{\partial \mathbf{A}}{\partial t} + \epsilon\frac{\partial^2 \mathbf{A}}{\partial t^2} + (1/\mu)[\nabla(\nabla.\mathbf{A}) - \nabla^2\mathbf{A}] \tag{1.30}$$

Dividing by σ

$$-\nabla\Psi - (\epsilon/\sigma)\frac{\partial\nabla\Psi}{\partial t} = \frac{\partial\mathbf{A}}{\partial t} + (\epsilon/\sigma)\frac{\partial^2\mathbf{A}}{\partial t^2} + (1/\mu\sigma)[\nabla(\nabla.\mathbf{A}) - \nabla^2\mathbf{A}] \quad (1.31)$$

Applying the Lorenz condition [34]

$$\nabla.\mathbf{A} = -\epsilon\mu\frac{\partial\Psi}{\partial t} \tag{1.32}$$

$$\nabla(\nabla.\mathbf{A}) = -\epsilon\mu\frac{\partial\nabla\Psi}{\partial t} \tag{1.33}$$

Substituting (1.33) into (1.31) gives

$$-\nabla\Psi - (\epsilon/\sigma)\frac{\partial\nabla\Psi}{\partial t} = \frac{\partial\mathbf{A}}{\partial t} + (\epsilon/\sigma)\frac{\partial^2\mathbf{A}}{\partial t^2} + (1/\mu\sigma)[-\epsilon\mu\frac{\partial\nabla\Psi}{\partial t} - \nabla^2\mathbf{A}] \quad (1.34)$$

$$-\nabla\Psi = (\epsilon/\sigma)\frac{\partial^2\mathbf{A}}{\partial t^2} + \frac{\partial\mathbf{A}}{\partial t} - (1/\mu\sigma)\nabla^2\mathbf{A} \tag{1.35}$$

Thus since $-\nabla\Psi = V_{emf}$ then the general equation for the *emf* in the time domain is

$$V_{emf} = (\epsilon/\sigma)\frac{\partial^2\mathbf{A}}{\partial t^2} + \frac{\partial\mathbf{A}}{\partial t} - (1/\mu\sigma)\nabla^2\mathbf{A} \tag{1.36}$$

For cylindrical co-ordinates this becomes

$$V_{emf} = -(1/\mu\sigma)\left[\frac{\partial^2\mathbf{A}}{\partial r^2} + \frac{\partial\mathbf{A}}{r\partial r} - \mu\epsilon\frac{\partial^2\mathbf{A}}{\partial t^2} - \mu\sigma\frac{\partial\mathbf{A}}{\partial t}\right] \tag{1.37}$$

Multiplying (1.35) by σ gives

$$-\sigma\nabla\Psi = \epsilon\frac{\partial^2\mathbf{A}}{\partial t^2} + \sigma\frac{\partial\mathbf{A}}{\partial t} - (1/\mu)\nabla^2\mathbf{A} \tag{1.38}$$

Re-arranging (1.35) we obtain

$$-\nabla\Psi - \frac{\partial\mathbf{A}}{\partial t} = (\epsilon/\sigma)\frac{\partial^2\mathbf{A}}{\partial t^2} - (1/\mu\sigma)\nabla^2\mathbf{A} = \mathbf{E} \tag{1.39}$$

That is

$$\nabla^2\mathbf{A} - \mu\epsilon\frac{\partial^2\mathbf{A}}{\partial t^2} = -\mu\sigma\mathbf{E} = -\mu\mathbf{J} \tag{1.40}$$

A similar equation may be obtained for the scalar potential Ψ. From (1.3) and Maxwell's equation $\nabla.\mathbf{E} = \rho_v/\epsilon$

$$\nabla.\mathbf{E} = -\frac{\partial\nabla.\mathbf{A}}{\partial t} - \nabla.(\nabla\Psi) = \rho_v/\epsilon \tag{1.41}$$

Applying the Lorenz condition (1.32) gives

$$\nabla^2\Psi - \mu\epsilon\frac{\partial^2\Psi}{\partial t^2} = -\frac{\rho_v}{\epsilon} \tag{1.42}$$

Equations (1.40) and (1.42) are inhomogeneous wave equations for vector and scalar potentials respectfully. Their solutions are discussed in detail elsewhere [22,43].

1.6.1   Non-conductors

For materials with very high resistivity and conductivity close to zero, that is, $\sigma \approx 0$ then (1.40) becomes

$$\nabla^2 \mathbf{A} - \mu\epsilon \frac{\partial^2 \mathbf{A}}{\partial t^2} = 0 \tag{1.43}$$

which is a homogeneous wave equation agreeing with Maxwell, (1.16).

1.6.2   Conductors

For materials with very high conductivity and negligible displacement current, then $(\epsilon/\sigma) \approx 0$. In copper for example $(\epsilon/\sigma) \approx 10^{-20}$. Then from (1.39) or (1.40)

$$\nabla^2 \mathbf{A} = -\mu\sigma \mathbf{E} = -\mu\mathbf{J} \tag{1.44}$$

Equation (1.39) also gives

$$-\nabla\Psi - \frac{\partial \mathbf{A}}{\partial t} = -(1/\mu\sigma)\nabla^2 \mathbf{A} \tag{1.45}$$

This may also be re-written in terms of the emf since $V_{emf} = -\nabla\Psi$. Hence,

$$V_{emf} = -\nabla\Psi = \frac{\partial \mathbf{A}}{\partial t} - (1/\mu\sigma)\nabla^2 \mathbf{A} \tag{1.46}$$

These equations are used in the time domain analysis of conductors [82]. The impedance of the conductor in the time domain is

$$Z = \frac{V_{emf}}{I} = \frac{V_{emf}}{\int_S \mathbf{J}.d\mathbf{S}} = -(1/\mu\sigma)\nabla^2 \mathbf{A} - (1/\mu)\int_S \nabla^2 \mathbf{A} d\mathbf{S} = -\frac{[\mu\frac{\partial \mathbf{A}}{\partial t} - \rho\nabla^2 \mathbf{A}]}{\int_S \nabla^2 \mathbf{A} d\mathbf{S}} \tag{1.47}$$

Substituting for (1.44) leads to

$$Z = \frac{\frac{\partial \mathbf{A}}{\partial t} + \rho\mathbf{J}}{\int_S \mathbf{J}.d\mathbf{S}} = \left(\mathbf{E} + \frac{\partial \mathbf{A}}{\partial t}\right)/I = \frac{V_{emf}}{I} \tag{1.48}$$

agreeing with first part of equation (1.47). For conduction in the z-direction only, Faraday's Law gives for the time dependent emf

$$V_{emf}(t) = -L_{ext}dI/dt = -dA_z/dt \tag{1.49}$$

where L_{ext} is the external inductance. The total *emf* is then

$$V_{emf} = E_z + L_{ext}dI/dt \tag{1.50}$$

The impedance is

$$Z = V_{emf}/I = (\rho J_z + L_{ext}dI/dt)/I \tag{1.51}$$

For current flowing in the z-direction only in a solid cylindrical conductor then (1.46) becomes

$$V_{emf} = -\frac{\rho l}{\mu}\left(\frac{\partial^2 A_z}{\partial r^2} + \frac{1}{r}\frac{\partial A_z}{\partial r}\right) + l\frac{\partial A_z}{\partial t} \tag{1.52}$$

Maxwell [3,17,21] Art. 689 solved this equation by assuming A_z to be represented by the series

$$A_z = S + T_o + T_1 r^2 + T_2 r^4 + T_3 r^6 + \cdots + T_n r^{2n} + \cdots + \tag{1.53}$$

where S, T_o, T_1, etc. are functions of time. After differentiating equation (1.53) and after some considerable algebra [36] or using matrix algebra [66] the *emf* becomes

$$V_{emf} = R_o I + L_o I^{(1)} - \frac{\mu^2 l^2}{12 R_o} I^{(2)} + \frac{\mu^3 l^3}{48 R_o^2} I^{(3)} - \frac{\mu^4 l^4}{180 R_o^3} I^{(4)} +, \text{etc.} \tag{1.54}$$

where $\mu = \mu_o \mu_r / (4\pi)$ and $I^{(m)}$ is the mth derivative of the current. I is the current, R_o and L_o are the low frequency resistance and inductance, respectively. This inductance is the sum of the internal inductance of the conductor plus its external inductance, that is, $L_o = L_{int} + L_{ext}$ where it is derived in the vector potential approach rather than assumed, [82]. Equation (1.54) may be re-written as

$$V_{emf} = R_o I + L_o \frac{dI}{dt} - R_o \sum_{m=2}^{\infty} (-1)^m f_m \left(\frac{a}{d}\right)^{2m} I^{(m)} \tag{1.55}$$

where

$$d = \sqrt{\frac{4\rho}{\mu_o}} \text{ ms}^{-1/2}, \quad I^{(m)} = \frac{d^m I}{dt^m}, \quad f_2 = \frac{1}{12}, \quad f_3 = \frac{1}{48}, \quad f_4 = \frac{1}{180} \ldots, \text{ etc.} \tag{1.56}$$

Previously, the *f*-coefficients were found to fit a power law relationship [82]

$$f_m = a_r 10^{b_r m}, \quad m = 1, 2, 3, 4, 5, \text{etc.} \tag{1.57}$$

where $a_r = 1.13721$ and $b_r = -0.5675$.

Maxwell's approach to solving the vector potential diffusion equation for current flowing in a long cylindrical conductor leads to a power series 1.55. This gives the correct results at low frequencies and is consistent with Bessel function solutions. However, it turns out that the series solution diverges if the ratio of conductor radius to skin depth a/δ exceeds 2.7. This is close to the mathematical irrational number $e = 2.718$. This divergence has been considered previously to be a consequence of mathematical inversion, but this has not yet been proved for the power series case [82]. The problem of series solution instability is also discussed by Arfken [6]. Previously, there has been criticism of Maxwell's analysis of current flowing in a 'solitary wire' [39], and this has prevented publication of some papers on this topic. However, it turned out these criticisms were invalid since in Art. 682 Maxwell stated explicitly that he was referring to a complete circuit of 'two very long parallel conductors' [82]. But it has also been shown that the equations also apply to a solitary conductor [40].

1.6.3   Solution using Bessel functions

For sine waves the current density in terms of Bessel functions is given by [38]

$$J_z = J_{dc} \frac{u_a}{2} \frac{J_0(u)}{J_1(u_a)} \tag{1.58}$$

and the impedance is

$$Z = R_{dc}\frac{u_a}{2}\frac{J_0(u)}{J_1(u_a)} + j\omega L_{ext} \tag{1.59}$$

where $J_o(u)$ and $J_1(u_a)$ are zero and first order Bessel functions of the first kind respectively,

$$R_{dc} = \frac{\rho_z}{\pi a^2} \tag{1.60}$$

$$u = j^{3/2}\sqrt{2}(r/\delta), \quad u_a = j^{3/2}\sqrt{2}(a/\delta) \tag{1.61}$$

$$\delta = \sqrt{2\rho/\mu\omega} \tag{1.62}$$

and δ is the skin depth. It is interesting that Maxwell solved the vector potential non-homogeneous partial differential (1.52), which includes the source, whereas most authors solve homogeneous partial differential equations (PDEs) in terms of the electric and magnetic fields, which exclude the source [1,12,38,77].

1.7   Final equations for the time dependent electromagnetic field

From (1.38)

$$\nabla^2\mathbf{A} = \mu\epsilon\frac{\partial^2\mathbf{A}}{\partial t^2} + \mu\sigma\frac{\partial\mathbf{A}}{\partial t} + \mu\sigma\nabla\Psi \tag{1.63}$$

and from the *Helmholtz Wave Equations* 2.50 and 2.54

$$\nabla^2\mathbf{E} = \mu\epsilon\frac{\partial^2\mathbf{E}}{\partial t^2} + \mu\sigma\frac{\partial\mathbf{E}}{\partial t} + \nabla(\rho_v/\epsilon) \tag{1.64}$$

$$\nabla^2\mathbf{H} = \mu\epsilon\frac{\partial^2\mathbf{H}}{\partial t^2} + \mu\sigma\frac{\partial\mathbf{H}}{\partial t} \tag{1.65}$$

1.8   Summary and discussion

In this chapter, I have reviewed James Clerk Maxwell's theory of electromagnetic disturbance through a uniform medium. Maxwell's approach was to use vector potentials. This is in contrast to the common method, which solves the electric and magnetic fields separately using 'Maxwell's equations' to formulate the Helmholtz equations, then setting the diffusion terms and the charge density to zero. In this, I have compared Maxwell's expressions, which used old German Euler Fracture fonts and electromagnetic units, with modern expressions that use SI units. Apart from a few differences in expressions for the components of the vector potential F, G, H and negative Laplacian $(-\nabla^2)$ used by Maxwell, the modern equations, as expected, are the same as Maxwell's. In deriving the wave equations for propagation in non-conductors, Maxwell effectively used the Coulomb Gauge by setting $J_M = \nabla.\mathbf{A} = 0$ although he does not describe this as 'Coulomb Gauge'.

In the alternative approach to obtaining the general equations of electromagnetic disturbances, the Lorentz condition (1.32) is applied, which yields source-dependent inhomogeneous wave and diffusion equations for the vector potential. The impedance is then determined for the time domain in terms of the vector potential only equation (1.47) and the frequency domain using Bessel functions equation (1.59).

1.9   Appendix

1.9.1   Proof of (1.3)

From Maxwell's equation for Faraday's law

$$\nabla \times \mathbf{E} = -\frac{\partial \mathbf{B}}{\partial t} = -\frac{\partial}{\partial t}\nabla \times \mathbf{A} \tag{1.66}$$

or

$$\nabla \times \left(\mathbf{E} + \frac{\partial \mathbf{A}}{\partial t}\right) = 0 \tag{1.67}$$

The vector quantity in (1.67) has no curl, that is, it is irrotational and can be derived from the gradient of a scalar potential, $-\nabla\Psi$. Hence,

$$\mathbf{E} = -\nabla\Psi - \frac{\partial \mathbf{A}}{\partial t} \tag{1.68}$$

which is proof of (1.3). This equation essentially expresses Helmholtz's theorem, which is more rigorously proved in reference [34].

An alternative method [20] applies Faraday's law via the magnetic flux Φ. For a closed surface S

$$\Phi = \int_S \mathbf{B}.d\mathbf{S} = \int_S \nabla \times \mathbf{A}.d\mathbf{S} = \oint_l \mathbf{A}.d\mathbf{l} \tag{1.69}$$

Faraday's Law gives the *emf* as

$$V_{emf} = \oint_l \mathbf{E}.d\mathbf{l} = -\frac{\partial \Phi}{\partial t} = -\oint_l \frac{\partial \mathbf{A}}{\partial t}.d\mathbf{l} \tag{1.70}$$

Hence,

$$\mathbf{E} = -\frac{\partial \mathbf{A}}{\partial t} \tag{1.71}$$

For steady state fields (d.c.), the electric field is given by the negative gradient of a scalar potential only i.e.

$$\mathbf{E} = -\nabla\Psi \tag{1.72}$$

The total field is then

$$\mathbf{E} = -\nabla\Psi - \frac{\partial \mathbf{A}}{\partial t} \tag{1.73}$$

The first approach assumes the most general electric field, which has both non-zero divergence and non-zero curl. This can be derived from the negative gradient of a scalar potential $-\nabla\Psi$ and a vector potential $\mathbf{A}$; a statement of Helmholtz's theorem. The second method finds the sum of steady-state and time-dependent fields. The two approaches lead to the same result (1.3).

1.9.2   Proof of (1.5)

From Maxwell Arts. 616 and 783, Equation (4),

$$\frac{\partial \gamma}{\partial y} = \left(\frac{\partial^2 G}{\partial x \partial y} - \frac{\partial^2 F}{\partial y^2} \right) / \mu_1 \tag{1.74}$$

$$\frac{\partial \beta}{\partial z} = \left(\frac{\partial^2 F}{\partial z^2} - \frac{\partial^2 H}{\partial x \partial z} \right) / \mu_1 \tag{1.75}$$

$$4\pi u \mu_1 = \frac{\partial \gamma}{\partial y} - \frac{\partial \beta}{\partial z} = \left(\frac{\partial^2 G}{\partial x \partial y} - \frac{\partial^2 F}{\partial y^2} - \frac{\partial^2 F}{\partial z^2} + \frac{\partial^2 H}{\partial x \partial z} \right) = Y \tag{1.76}$$

$$Y = \left(\frac{\partial^2 G}{\partial x \partial y} - \left(\frac{\partial^2 F}{\partial x^2} + \frac{\partial^2 F}{\partial y^2} + \frac{\partial^2 F}{\partial z^2} \right) + \frac{\partial^2 F}{\partial x^2} + \frac{\partial^2 H}{\partial x \partial z} \right) \tag{1.77}$$

$$\frac{\partial J_M}{\partial x} = \frac{\partial^2 F}{\partial x^2} + \frac{\partial^2 G}{\partial x \partial y} + \frac{\partial^2 H}{\partial x \partial z} \tag{1.78}$$

$$\nabla^2 F = - \left(\frac{\partial^2 F}{\partial x^2} + \frac{\partial^2 F}{\partial y^2} + \frac{\partial^2 F}{\partial z^2} \right) \tag{1.79}$$

Hence,

$$4\pi u \mu_1 = \frac{\partial J_M}{\partial x} + \nabla^2 F \tag{1.80}$$

Similarly,

$$4\pi v \mu_1 = \frac{\partial J_M}{\partial y} + \nabla^2 G \tag{1.81}$$

$$4\pi w \mu_1 = \frac{\partial J_M}{\partial z} + \nabla^2 H \tag{1.82}$$

1.9.3   Proof of the SI equation (1.5)

$$\mathbf{J} = \nabla \times \mathbf{H} = (1/\mu)\nabla \times \mathbf{B} = (1/\mu)\nabla \times \nabla \times \mathbf{A} = (1/\mu)[\nabla(\nabla.\mathbf{A} - \nabla^2\mathbf{A}] \tag{1.83}$$

Hence,

$$\mu\mathbf{J} = -\nabla^2\mathbf{A} + \nabla(\nabla.\mathbf{A}) \tag{1.84}$$

1.9.4   Alternative derivation of (1.29)

From the Helmholtz wave equation for the electric field (2.50)

$$\nabla^2\mathbf{E} = \mu\sigma \frac{\partial \mathbf{E}}{\partial t} + \mu\epsilon \frac{\partial^2 \mathbf{E}}{\partial t^2} + \nabla(\rho_v/\epsilon) \tag{1.85}$$

$$\nabla^2\mathbf{E} = \nabla(\nabla.\mathbf{E}) - \nabla \times \nabla \times \mathbf{E} \tag{1.86}$$

Using (1.3)

$$\mathbf{E} = -\frac{\partial \mathbf{A}}{\partial t} - \nabla \Psi \tag{1.87}$$

$$\nabla^2 \mathbf{E} = \nabla(\nabla.\mathbf{E}) - \nabla \times \nabla \times \mathbf{E}$$
$$= \mu\sigma \frac{\partial \left(-\frac{\partial \mathbf{A}}{\partial t} - \nabla\Psi \right)}{\partial t} + \mu\epsilon \frac{\partial^2 \left(-\frac{\partial \mathbf{A}}{\partial t} - \nabla\Psi \right)}{\partial t^2} + \nabla(\rho_v/\epsilon) \tag{1.88}$$

Now the ρ_v term cancels because

$$\nabla(\nabla.\mathbf{E}) = \nabla(\rho_v/\epsilon) \tag{1.89}$$

Also

$$\nabla \times \nabla \times \mathbf{E} = -\nabla \times \frac{\partial \mathbf{B}}{\partial t} = -\frac{\partial}{\partial t}\nabla \times \nabla \times \mathbf{A} \tag{1.90}$$

$$-\nabla \times \nabla \times \mathbf{E} = \frac{\partial}{\partial t}\nabla \times \nabla \times \mathbf{A}$$
$$= -\mu\sigma \frac{\partial^2 \mathbf{A}}{\partial t^2} - \mu\epsilon\frac{\partial^3 \mathbf{A}}{\partial t^3} - \mu\epsilon\frac{\partial^2 \nabla\Psi}{\partial t^2} - \mu\sigma\frac{\partial\nabla\Psi}{\partial t} \tag{1.91}$$

Integrating with respect to t

$$= (1/\mu)\nabla \times \nabla \times \mathbf{A} = -\left(\sigma\frac{\partial \mathbf{A}}{\partial t} + \epsilon\frac{\partial^2 \mathbf{A}}{\partial t^2} + \epsilon\frac{\partial\nabla\Psi}{\partial t} + \sigma\nabla\Psi \right) \tag{1.92}$$

$$\sigma\frac{\partial \mathbf{A}}{\partial t} + \epsilon\frac{\partial^2 \mathbf{A}}{\partial t^2} + \epsilon\frac{\partial\nabla\Psi}{\partial t} + \sigma\nabla\Psi + (1/\mu)(\nabla(\nabla.\mathbf{A} - \nabla^2\mathbf{A}) = 0 \tag{1.93}$$

which is the same as (1.29).

1.9.5   *The continuity equation*

The continuity equation arises from the consideration of charge conservation in analogy with the conservation of energy and the conservation of matter. Consider a closed surface S, volume V, containing charge +Q. The total current emerging is

$$I = \oint_S \mathbf{J}.d\mathbf{S} = -\frac{dQ}{dt} = -\frac{d}{dt}\int_V \rho_v dV \tag{1.94}$$

The divergence theorem is

$$\oint_S \mathbf{J}.d\mathbf{S} = \int_V \nabla.\mathbf{J}dV \tag{1.95}$$

Therefore

$$\nabla.\mathbf{J} = -\frac{\partial\rho_v}{\partial t} \tag{1.96}$$

Thus the current or charge per second diverging from a unit volume is equal to the time rate of decrease of charge per unit volume.

For finite conductivity, substitute for $\mathbf{J}$ from (1.1) and using Maxwell's first equation $\nabla.\mathbf{E} = \rho_v/\epsilon$ then

$$\nabla.\mathbf{J} = \nabla.\sigma\mathbf{E} + \epsilon\frac{\partial \nabla.\mathbf{E}}{\partial t} = \frac{\partial \rho_v}{\partial t} + \frac{\sigma\rho_v}{\epsilon} \tag{1.97}$$

Equating this with equation (1.96) gives

$$\frac{\partial \rho_v}{\partial t} + \frac{\sigma\rho_v}{2\epsilon} = 0 \tag{1.98}$$

with a solution

$$\rho_v = \rho_{v0}e^{-kt}, \quad k = \sigma/(2\epsilon) \tag{1.99}$$

where ρ_{v0} is the charge at $t = 0$. In the steady state (d.c.)

$$\frac{\partial \rho_v}{\partial t} = 0, \quad \nabla.\mathbf{J} = 0, \quad \rho_v = \rho_{v0} \tag{1.100}$$

Discharge of a Capacitor

The discharge of a capacitor through a resistor gives a similar result. This can be obtained using circuit theory. Summing the voltage across the capacitor V_C and resistor V_R, Kirchhoff's voltage law gives $V_C + V_R = 0$. The charge equation is then

$$\frac{q}{C} + R\frac{dq}{dt} = 0 \tag{1.101}$$

This has a solution

$$q = q_0e^{-t/(CR)} \tag{1.102}$$

where q_0 is the charge at $t = 0$. This is similar to (1.99) derived from the continuity equation. Note that in this analysis we have assumed ideal C and R components with no inductance in the circuit. For very rapid discharged the skin effect of the conductors may become significant. In this case R becomes dependent on the discharge time or frequency.

1.9.6 Proof that the Lorenz condition leads to the continuity equation

Taking Laplacian of the Lorenz equation (1.32)

$$\nabla^2\nabla.\mathbf{A} = \nabla.\nabla^2\mathbf{A} = -\epsilon\mu\frac{\partial \nabla^2\Psi}{\partial t} \tag{1.103}$$

Substitute for (1.40) and (1.42)

$$\nabla.\left(-\mu\mathbf{J} + \mu\epsilon\frac{\partial^2\mathbf{A}}{\partial t^2}\right) = -\epsilon\mu\frac{\partial}{\partial t}\left(\epsilon\mu\frac{\partial^2\Psi}{\partial t^2} - \rho_v/\epsilon\right) \tag{1.104}$$

$$-\mu\nabla.\mathbf{J} + \mu\epsilon\frac{\partial^2\nabla.\mathbf{A}}{\partial t^2} = -(\epsilon\mu)^2\frac{\partial^3\Psi}{\partial t^3} + \frac{\partial\mu\rho_v}{\partial t} \tag{1.105}$$

Applying Lorenz equation (1.32) to the the LHS of (1.105) then this becomes

$$-\mu\nabla.\mathbf{J} + \mu\epsilon\frac{\partial^2}{\partial t^2}\left(-\epsilon\mu\frac{\partial\Psi}{\partial t}\right) = -\mu\nabla.\mathbf{J} - (\epsilon\mu)^2\frac{\partial^3\Psi}{\partial t^3} \tag{1.106}$$

This is then equated with the RHS of (1.105) to give

$$-\mu \nabla .\mathbf{J} - (\epsilon\mu)^2 \frac{\partial^3 \Psi}{\partial t^3} = -(\epsilon\mu)^2 \frac{\partial^3 \Psi}{\partial t^3} + \frac{\partial \mu \rho_v}{\partial t} \tag{1.107}$$

This gives

$$\nabla .\mathbf{J} = -\frac{\partial \rho_v}{\partial t} \tag{1.108}$$

which is the continuity equation.

As an alternative proof, consider only time-dependent current flow as in the derivation of the continuity equation. Hence, from (1.1) and (1.3)

$$\nabla .\mathbf{J} = \epsilon \left(\frac{\partial^2 \nabla .\mathbf{A}}{\partial t^2} + \frac{\partial \nabla^2 \Psi}{\partial t} \right) \tag{1.109}$$

Substituting for $\nabla .\mathbf{A}$ from the Lorenz condition (1.32) gives

$$\nabla .\mathbf{J} = \epsilon \frac{\partial}{\partial t} \left(-\epsilon\mu \frac{\partial^2 \Psi}{\partial t^2} + \nabla^2 \Psi \right) \tag{1.110}$$

Now substitute for $-\rho_v/\epsilon$ from (1.42)

$$\nabla^2 \Psi - \mu\epsilon \frac{\partial^2 \Psi}{\partial t^2} = -\frac{\rho_v}{\epsilon} \tag{1.111}$$

Therefore

$$\nabla .\mathbf{J} = -\frac{\partial \rho_v}{\partial t} \tag{1.112}$$

which yields the fundamental continuity (1.96) obtained again using the Lorenz condition.

1.9.7 Proof of (1.37)

From (1.36) for cylindrical co-ordinates with radius r and the vector A in the z-direction only then

$$\nabla^2 A_z = \frac{1}{r} \frac{\partial}{\partial r} \left(r \frac{\partial A_z}{\partial r} \right) + \frac{1}{r^2} \frac{\partial^2 A_z}{\partial \phi^2} + \frac{\partial^2 A_z}{\partial z^2} \tag{1.113}$$

For $A_z = f(r, t)$ only and constant in the z-direction then

$$\nabla^2 A_z = \frac{1}{r} \frac{\partial}{\partial r} \left(r \frac{\partial A_z}{\partial r} \right) \tag{1.114}$$

Expanding this equation gives

$$\nabla^2 A_z = \frac{1}{r} \left(\frac{\partial A_z}{\partial r} + r \frac{\partial^2 A_z}{\partial r^2} \right) \tag{1.115}$$

Hence (1.36) becomes

$$V_{emf} = -(1/\mu\sigma) \left[\frac{\partial^2 \mathbf{A}}{\partial r^2} + \frac{\partial \mathbf{A}}{r \partial r} - \mu\epsilon \frac{\partial^2 \mathbf{A}}{\partial t^2} - \mu\sigma \frac{\partial \mathbf{A}}{\partial t} \right] \tag{1.116}$$

1.10   Table of Euler Fraktur fonts

A	B	C	D	E	F	G	H	I	J	K	L	M	N	O	P	Q	R	S	T	U	V	W	X	Y	Z
𝔄	𝔅	ℭ	𝔇	𝔈	𝔉	𝔊	ℌ	ℑ	𝔍	𝔎	𝔏	𝔐	𝔑	𝔒	𝔓	𝔔	ℜ	𝔖	𝔗	𝔘	𝔙	𝔚	𝔛	𝔜	ℨ

Chapter 2

Solution of Maxwell's equations in loss free and lossy media

2.1 Introduction

The theories of classical electromagnetism have been reduced to four equations – *Maxwell's Equations*. These four equations use vector calculus, considerably simplifying the original mathematics in Maxwell's Treatise [3]. This was proposed by Oliver Heaviside in 1884 and recently reviewed by D.P. Hampshire [8].

In the following, we assume a linear, isotropic medium where the path of the medium is stationary. This assumption of a stationary path applies to all the following chapters. In this case, Maxwell's four equations are [20]:

Differential form (fields)   Integral form (boundaries)

$$div\ \mathbf{D} = \nabla\cdot\mathbf{D} = \rho_v, \qquad \oint_S \mathbf{D}\cdot d\mathbf{S} = \int_{vol} \rho_V d(vol) \tag{2.1}$$

$$curl\ \mathbf{E} = \nabla\times\mathbf{E} = -\frac{\partial\mathbf{B}}{\partial t}, \qquad \oint_l \mathbf{E}\cdot d\mathbf{l} = -\frac{\partial}{\partial t}\int_S \mathbf{B}\cdot d\mathbf{S} \tag{2.2}$$

$$curl\ \mathbf{H} = \nabla\times\mathbf{H} = \mathbf{J} + \frac{\partial\mathbf{D}}{\partial t}, \qquad \oint_l \mathbf{H}\cdot d\mathbf{l} = I + \int_S \frac{\partial\mathbf{D}}{\partial t}\cdot d\mathbf{S} \tag{2.3}$$

$$div\ \mathbf{B} = \nabla\cdot\mathbf{B} = 0, \qquad \oint_S \mathbf{B}\cdot d\mathbf{S} = 0 \tag{2.4}$$

where the electric field $\mathbf{E}$, electric flux $\mathbf{D} = \epsilon\mathbf{E}$, current density $\mathbf{J} = \sigma\mathbf{E}$, magnetic flux density $\mathbf{B} = \mu\mathbf{H}$, ρ_v is the volume charge density, σ is the material electrical conductivity. ϵ_r is the relative permittivity and χ_e the electric susceptibility.

The original experiments and theories for these equations were due to:

- Equation (2.1) Faraday's experiments on electric flux and Gauss's theorem.
- Equation (2.2) Faraday's experiments and Law.
- Equation (2.3) Ampere's law, Biot and Savart's experiments plus Maxwell's displacement current.
- Equation (2.4) Gauss's law for magnetic fields or Maxwell's equation [3], Eq. 17, Art. 402.

In the above equations

$$\epsilon = \epsilon_o \epsilon_r = \epsilon_o (1 + \chi_e), \quad \mu = \mu_o \mu_r = \mu_o (1 + \chi_m) \tag{2.5}$$

where μ_r is the relative permeability and χ_m is the magnetic susceptibility. These parameters arise from electric and magnetic polarisation of the atoms or molecules in the material. This polarisation has a time dependence that leads to hysteresis between the applied fields and the fields in the material. These effects can be accounted for by expressing the parameters in complex form:

$$\epsilon = \epsilon' - j\epsilon'', \quad \mu = \mu' - j\mu'', \quad \sigma = \sigma' - j\sigma'' \tag{2.6}$$

where the real and imaginary components are indicated by the prime (*/*) and double prime (*//*), respectively, this hysteresis leads to energy loss due to heat dissipation. This loss is added to the conductive loss due to the material conductivity σ. The negative signs arise because in a passive material, such as considered here, there can be no energy gain.

The complex ohmic conductivity arises because electron scattering takes place in a specific time; the *electron relaxation time*, τ. Hence, at frequencies that approach $1/\tau$, the conductivity becomes complex. The high-frequency conductivity is given by Chambers [72]

$$\sigma_{hf} = \frac{\sigma_o}{1 + j\omega\tau} = \frac{\sigma_o}{1 + (\omega\tau)^2}(1 - j\omega\tau) \tag{2.7}$$

where $\sigma_o = (nq^2\tau/m)$ is the low-frequency conductivity, n the electron density, q the charge on the electron and m the electron mass.

The current flow is in the same direction as the applied field, but it is shifted in phase by an angle $\theta = tan^{-1}(-\omega\tau)$. If $\omega\tau \gg 1$, $\sigma_{hf} = -j\sigma_o/(\omega\tau) = nq^2/(dm/dt)$ where $dm/dt = j\omega m$. Hence, the high-frequency conductivity depends on the mass inertia of the electrons. To estimate the frequency at which this occurs, we require a value for the relaxation time. For copper $\tau_{Cu} = 2.4 \times 10^{-14}$ s. For the high-frequency condition, we require $\omega \gg 1/\tau$ or $f \gg 6.4 \times 10^{12}$ Hz. This corresponds to frequencies in the far infrared region of the electromagnetic spectrum.

If the material is anisotropic, then σ, ϵ and μ may depend on direction in the material. Each of these parameters can then be expressed as tensors. Polycrystalline metals, such as copper, silver, gold and aluminium, are isotropic. Semiconductor crystals Si, Ge, GaAs, etc., are anisotropic. Also, ferrite materials used in microwave circuits are anisotropic. Isotropic materials are considered in the following unless mentioned otherwise. The applied fields are also assumed to be sinusoidal so that the time derivative can be replaced by $j\omega$ where ω. is the angular frequency. If the material has electrical properties that are frequency-dependent, that is, dispersive, then it may be possible to transform the time-dependent field into a series of sinusoidal waves using Fourier analysis. The following analysis would then apply only to a single frequency say ω_1.

2.2   Solution of Maxwell's equations in free space

Maxwell's equations for vacuum or free space are:

$$div\ \mathbf{D} = \rho_v = 0 \tag{2.8}$$

$$curl\ \mathbf{E} = -\frac{\partial \mathbf{B}}{\partial t} = -\mu_o \frac{\partial \mathbf{H}}{\partial t} \tag{2.9}$$

$$curl\ \mathbf{H} = \mathbf{J} + \frac{\partial \mathbf{D}}{\partial t} = \epsilon_o \frac{\partial \mathbf{E}}{\partial t} \tag{2.10}$$

$$div\ \mathbf{B} = 0 \tag{2.11}$$

where for vacuum $\epsilon_r = 1$ and $\mathbf{D} = \epsilon_o \mathbf{E}$. Hence, $div\ \mathbf{D} = div\ \epsilon_o \mathbf{E} = \rho_v = 0$. Since there are no free charges present, the electric field cannot terminate on charges, but this does not mean that the electric field does not exist. The Maxwell curl equations lead to a time rate of change in $\mathbf{E}$ and $\mathbf{H}$ fields. Although the steady state current density $\mathbf{J}$ is zero, the time rate of change of $\mathbf{E}$ corresponds to a displacement current density. We now show that the solution of Maxwell's equations lead to wave motion of the electric and magnetic fields in free space.

2.3   Wave Motion in free space

From Maxwell's equations, taking the *curl(curl* $\mathbf{E}$)

$$curl(curl\ \mathbf{E}) = -\mu_o \frac{\partial(curl\ \mathbf{H})}{\partial t} = -\mu_o \frac{\partial}{\partial t}\left[\epsilon \frac{\partial \mathbf{E}}{\partial t}\right] \tag{2.12}$$

or

$$\nabla \times \nabla \times \mathbf{E} + \mu_o \epsilon_o \frac{\partial^2 \mathbf{E}}{\partial t^2} = 0 \tag{2.13}$$

But

$$\nabla \times \nabla \times \mathbf{E} = \nabla(\nabla . \mathbf{E}) - \nabla^2 \mathbf{E} \tag{2.14}$$

From Maxwell's first equation (2.8), $\nabla . \mathbf{E} = 0$. Hence,

$$\nabla^2 \mathbf{E} = \mu_o \epsilon_o \frac{\partial^2 \mathbf{E}}{\partial t^2} \tag{2.15}$$

Similarly, we can show that

$$\nabla^2 \mathbf{H} = \mu_o \epsilon_o \frac{\partial^2 \mathbf{H}}{\partial t^2} \tag{2.16}$$

The equation that represents a periodic wave travelling in the $+z$-direction is given by

$$\frac{\partial^2 f}{\partial z^2} = (1/u^2)\frac{\partial^2 f}{\partial t^2} \tag{2.17}$$

where f is the amplitude and u the velocity of a fixed point on the wave, that is, the phase velocity. This is shown in the following section. Comparing (2.25) with

(2.15) and (2.16), these equations represent periodic waves travelling at a phase velocity

$$u_o = \sqrt{\frac{1}{\mu_o \epsilon_o}}$$ (2.18)

By substituting values for the dielectric and magnetic constants, Maxwell found that u_o was close to the value of the velocity of light, c, measured at that time. Subsequent measurements have confirmed the theory that light consists of electromagnetic waves. The presently accepted experimentally measured velocity of light in vacuo is $c = 2.99\ 792\ 458 \times 10^8$ ms^{-1}. Since by definition $\mu_o = 4\pi \times 10^{-7}$ Hm^{-1}, then the defined value of $\epsilon_o = 1/(\mu_o c) = 8.854\ 187\ 817 \times 10^{-12}$ Fm^{-1}.

For sinusoidal waves, we can put $\frac{\partial}{\partial t} = j\omega$ and $\frac{\partial^2}{\partial t^2} = -\omega^2$. The wave equations, also known as the Helmholtz equations for the electric and magnetic fields, are then

$$\nabla^2 \mathbf{E} = -\beta^2 \mathbf{E}, \quad \nabla^2 \mathbf{H} = -\beta^2 \mathbf{H}$$ (2.19)

where β is the *phase constant* also referred to as the *wave number* k_o. Hence,

$$\beta = k_o = \omega \sqrt{\mu_o \epsilon_o} = \omega/c$$ (2.20)

Since $\omega = 2\pi f$ and $c = f\lambda_o$ where f is the frequency and λ_o is the wavelength of the EM waves in free space then $\beta = k_o = \frac{2\pi}{\lambda_o}$. Hence,

$$\lambda_o = \frac{2\pi}{\beta} = \frac{u_o}{f} \quad m$$ (2.21)

The refractive index is defined by $n = c/u = 1$ in free space.

2.3.1   Travelling waves

To show that (2.15) and (2.16) represent periodic waves consider an arbitrary wave propagating along the z-axis with phase velocity u and amplitude A, Figure 2.1. After a time t, a point on the wave has moved a distance ut. Assuming the amplitude is unchanged, then

$$A = f(z - ut) = f(ut - z)$$ (2.22)

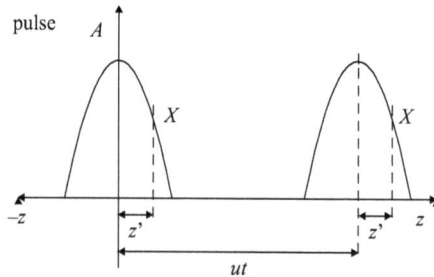

Figure 2.1   Wave travelling in the positive z-direction

represents a wave travelling in the +z-direction. Similarly, $A = f(ut + z)$ represents a wave travelling in the −z-direction. Taking the first and second derivatives with respect to distance gives

$$\frac{\partial A}{\partial z} = f'(ut - z), \quad \frac{\partial^2 A}{\partial z^2} = f''(ut - z) \tag{2.23}$$

Taking the first and second derivatives with respect to time gives

$$\frac{\partial A}{\partial t} = -uf'(ut - z), \quad \frac{\partial^2 A}{\partial t^2} = u^2 f''(ut - z) \tag{2.24}$$

where f' and f'' represent the first and second derivatives of the function f, respectively. Hence,

$$\frac{\partial^2 A}{\partial z^2} = u^{-2}\frac{\partial^2 A}{\partial t^2} \tag{2.25}$$

which is the wave equation for the arbitrary periodic pulse. If the wave is sinusoidal travelling in the positive z-direction a solution to this equation is

$$A = A_0 expj(\omega t - kz) \tag{2.26}$$

where $\omega = 2\pi f, f$ is the frequency, $k = \beta = 2\pi/\lambda$, where k is the wavenumber, β the phase constant and λ the wavelength. Differentiating twice with distance z and time t, we obtain the phase velocity $u = \omega/k$. If the wave is travelling in three dimensions then we can replace kz by $\mathbf{k.r}$ where $\mathbf{k}$ is the wave vector $\mathbf{k} = k_x\mathbf{i} + k_y\mathbf{j} + k_z\mathbf{k}$ and $\mathbf{r}$ is the radial distance vector $\mathbf{r} = x\mathbf{i} + y\mathbf{j} + z\mathbf{k}$. Hence,

$$A = A_0 expj(\omega t - \mathbf{k.r}) \tag{2.27}$$

2.3.2   Sinusoidal and non-sinusoidal waves

The example of sinusoidal waves previously discussed is of general practical importance. Non-sinusoidal waves can also be considered since any function of time, periodic or non-periodic, can be modelled by a spectrum of waves as in Fourier analysis. In this case, each sinusoidal or monochromatic wave can be represented by the complex or phasor form

$$\mathbf{E}_s = |\mathbf{E}|e^{j\omega t} = |\mathbf{E}|[cos(\omega t) + jsin(\omega t)] \tag{2.28}$$

the subscript 's' referring to a sinusoidal wave in the $s = j\omega$ form. The wave description can then be simplified by taking either the real (cosine) or imaginary (sine) component of the wave. Standard boldface form $\mathbf{E}$ etc is used in this book to avoid too much complexity similar to Plonsey and Collin [34]. The amplitude E (V/m) can be represented either by the maximum or peak value E_m or the root mean square $E_{rms} = E_m/\sqrt{2}$.

2.3.3   Relationship between E and H fields – TEM waves

For the magnetic field, expanding *curl* $\mathbf{E}$ in Maxwell's second equation (2.2),

$$\nabla \times \mathbf{E} = \left(\frac{\partial E_z}{\partial y} - \frac{\partial E_y}{\partial z}\right)\mathbf{a}_x + \left(\frac{\partial E_x}{\partial z} - \frac{\partial E_z}{\partial x}\right)\mathbf{a}_y + \left(\frac{\partial E_y}{\partial x} - \frac{\partial E_x}{\partial y}\right)\mathbf{a}_z \tag{2.29}$$

$$= \frac{\partial E_x}{\partial z}\mathbf{a}_y - \frac{\partial E_x}{\partial y}\mathbf{a}_z \tag{2.30}$$

Assuming $E_y = E_z = 0$, and there is no variation of E_x with y, then

$$\frac{\partial E_x}{\partial z}\mathbf{a}_y = -j\omega\mu_o\mathbf{H} \tag{2.31}$$

$$\mathbf{H} = -\frac{1}{j\omega\mu_o}\frac{\partial E_x}{\partial z}\mathbf{a}_y \tag{2.32}$$

But, $\frac{\partial E_x}{\partial z} = -\beta E_x$. Hence,

$$\mathbf{H} = \frac{\beta}{\omega\mu_o}E_x\mathbf{a}_y \quad \text{or} \quad H_y = \sqrt{\frac{\epsilon_o}{\mu_o}}E_{ox}e^{j(\omega t - \beta z)} \tag{2.33}$$

The *Intrinsic Impedance of Free Space* is then defined by

$$Z_o = \frac{E_x}{H_y} = \sqrt{\frac{\mu_o}{\epsilon_o}} = 376.73 \quad Ohms \tag{2.34}$$

Equation (2.33) indicates that the magnetic and electric fields are orthogonal or transverse to the direction of propagation, as shown in Figure 2.2. This is referred to as a *Transverse Electromagnetic wave*, abbreviated to 'TEM' wave. Note that the wave varies sinusoidally both in time and space. The wave is *polarised* because E and H have values in one direction only; x and y, respectively. This wave is also referred to as a *plane wave* because the electric and magnetic fields each occur only in a single plane, as shown in Figure 2.2. By definition, the direction of polarisation refers to the electric field direction. In the above case, the wave is polarised in the x-direction. Ordinary daylight is unpolarised, and the E vector moves in random directions due to sunlight being scattered by the atmosphere.

Waves propagated in waveguides can either be *tranverse electric* (TE) or *transverse magnetic* (TM) waves.

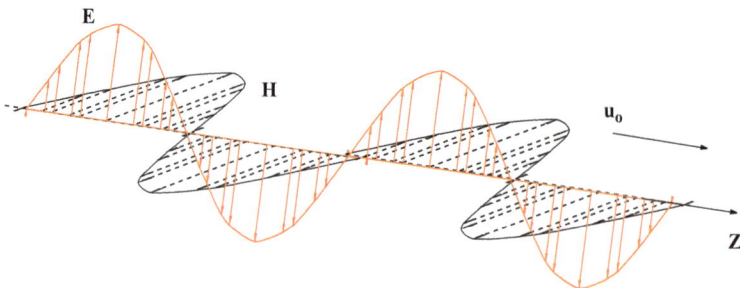

Figure 2.2   TEM plane wave propagating in free space showing the orthogonal E *and* H *fields*

2.3.3.1   TEM waves – example

Consider a plane wave with x-polarised electric vector $\mathbf{E} = E_x \mathbf{a}_x$ and y-polarised magnetic vector $\mathbf{H} = H_y \mathbf{a}_y$. The wave equations are

$$\frac{\partial^2 E_x}{\partial z^2} = -\beta^2 E_x, \quad \frac{\partial^2 H_y}{\partial z^2} = -\beta^2 H_y \tag{2.35}$$

The solutions for forward travelling waves are

$$E_x = E_1 e^{-j\beta z}, \quad H_y = H_1 e^{-j\beta z} \tag{2.36}$$

where

$$E_1 = E_o e^{j\omega t}, \quad H_1 = H_o e^{j\omega t} \tag{2.37}$$

To check this: $\frac{\partial E_x}{\partial z} = -j\beta E_1 e^{-j\beta z}$ and $\frac{\partial^2 E_x}{\partial z^2} = -\beta^2 E_1 e^{-j\beta z}$, that is, $\frac{\partial^2 E_x}{\partial z^2} = -\beta^2 E_x$ qed.

The time dependent solutions are then

$$E_x(z,t) = E_o e^{j(\omega t - \beta z)}, \quad H_y(z,t) = H_o e^{j(\omega t - \beta z)} \tag{2.38}$$

2.3.4   Plane wave with two components

Consider a free space plane EM wave propagated in the z-direction. The wave has two electric field components E_x and E_y, that is,

$$\mathbf{E} = E_x \mathbf{a}_x + E_y \mathbf{a}_y \tag{2.39}$$

The magnetic field is then

$$\mathbf{H} = -(E_y/Z_o)\mathbf{a}_x + (E_x/Z_o)\mathbf{a}_y \tag{2.40}$$

For forward waves with phase difference ϕ between E_x and E_y

$$\mathbf{E}(z,t) = E_{ox} e^{j(\omega t - \beta z)}\mathbf{a}_x + E_{oy} e^{j(\omega t - \beta z - \phi)}\mathbf{a}_y \tag{2.41}$$

$$\mathbf{H}(z,t) = -(E_{oy}/Z_o)e^{j(\omega t - \beta z - \phi)}\mathbf{a}_x + (E_{ox}/Z_o)e^{j(\omega t - \beta z)}\mathbf{a}_y \tag{2.42}$$

The phase difference ϕ and the relative magnitudes of E_x and E_y determines the type of polarisation of the wave as shown in Figure 2.3.

2.3.5   Free space propagation constants

In the case this case the medium has permittivity $\epsilon = \epsilon_o \epsilon_r$ and permeability $\mu = \mu_o \mu_r$. ϵ_r and μ_r are the relative permittivity and relative permeability respectively. The free space propagation constants now become

$$u = \sqrt{\frac{1}{\mu\epsilon}} = c/\sqrt{\mu_r \epsilon_r} \tag{2.43}$$

$$\beta = k = \omega\sqrt{\mu\epsilon} = (\omega/c)\sqrt{\mu_r \epsilon_r} \tag{2.44}$$

$$\lambda = \frac{2\pi}{\beta} = \frac{\lambda_o}{\sqrt{\mu_r \epsilon_r}} \tag{2.45}$$

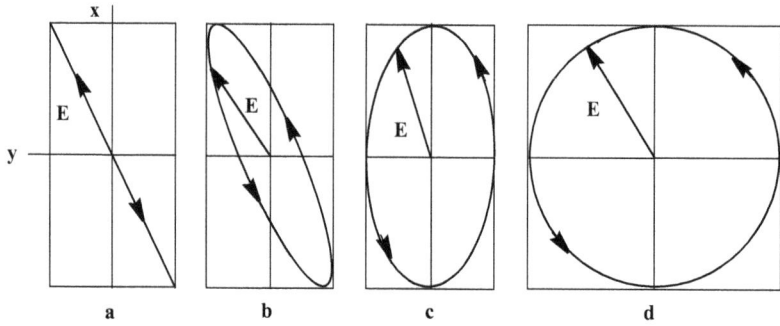

Figure 2.3   Plane wave with two components with $z = 0$: (a) linear polarisation: $E_x = 2, E_y = 1, \phi = 0$, (b) elliptical polarisation: $E_x = 2, E_y = 1, \phi = \pi/4$, (c) elliptical polarisation: $E_x = 2, E_y = 1, \phi = \pi/2$, (d) circular polarisation: $E_x = 2, E_y = 2, \phi = \pi/2$

$$Z = \frac{E_x}{H_y} = \sqrt{\frac{\mu}{\epsilon}} = Z_o\sqrt{\frac{\mu_r}{\epsilon_r}} \tag{2.46}$$

The refractive index is $n = c/u = \sqrt{\mu_r\epsilon_r}$. For dielectrics with $\mu_r = 1$ then $n = \sqrt{\epsilon_r}$.

2.4   Solution of Maxwell's equations-lossy medium

In this case Maxwell's equations are solved for a medium which has losses due to power dissipation. This may be due to ohmic resistance in conductors and hysteresis in ferrous conductors or leakage resistance and hysteresis in dielectrics. Hence, referring to the beginning of this chapter, taking the *curl(curl* **E**)

$$curl(curl\ \mathbf{E}) = -\mu\frac{\partial(curl\ \mathbf{H})}{\partial t} = -\mu\frac{\partial}{\partial t}\left[\sigma\mathbf{E} + \epsilon\frac{\partial\mathbf{E}}{\partial t}\right] \tag{2.47}$$

or

$$\nabla \times \nabla \times \mathbf{E} + \mu\sigma\frac{\partial\mathbf{E}}{\partial t} + \mu\epsilon\frac{\partial^2\mathbf{E}}{\partial t^2} = 0 \tag{2.48}$$

But

$$\nabla \times \nabla \times \mathbf{E} = \nabla(\nabla.\mathbf{E}) - \nabla^2\mathbf{E} \tag{2.49}$$

From Maxwell's first equation (2.1), $\nabla.\mathbf{E} = \rho_v/\epsilon$. Hence,

$$\nabla^2\mathbf{E} = \mu\sigma\frac{\partial\mathbf{E}}{\partial t} + \mu\epsilon\frac{\partial^2\mathbf{E}}{\partial t^2} + \nabla(\rho_v/\epsilon) \tag{2.50}$$

The equation that represents a periodic wave travelling in the +z-direction is given by

$$\frac{\partial^2 f}{\partial z^2} = (1/u^2)\frac{\partial^2 f}{\partial t^2} \tag{2.51}$$

where f is the amplitude and u the velocity of a fixed point on the wave, that is, the phase velocity. Comparing this with (2.50), then this represents a modified form of a simple wave which, in addition to the periodic term represented by the second derivative, includes a diffusion term and a term due to the gradient of the charge density.

For the **H** field, we have

$$\nabla \times \nabla \times \mathbf{H} = \nabla \times \mathbf{J} + \nabla \times \frac{\partial \mathbf{D}}{\partial t} = \nabla(\nabla.\mathbf{H}) - \nabla^2\mathbf{H} \tag{2.52}$$

From Maxwell's fourth equation (2.4), $\nabla.\mathbf{H} = 0$ then,

$$-\nabla^2\mathbf{H} = \sigma\nabla \times \mathbf{E} + \epsilon\nabla \times \frac{\partial \mathbf{E}}{\partial t} \tag{2.53}$$

Using Maxwell's second equation (2.2), then

$$\nabla^2\mathbf{H} = \mu\sigma\frac{\partial \mathbf{H}}{\partial t} + \mu\epsilon\frac{\partial^2 \mathbf{H}}{\partial t^2} \tag{2.54}$$

Equations (2.50) and (2.54) are the *Helmholtz Wave Equations* for time dependent electric and magnetic fields, respectively, in a lossy material also known as *The Equations of Telegraphy*.

For sinusoidal waves we can put $\frac{\partial}{\partial t} = j\omega$ and $\frac{\partial^2}{\partial t^2} = -\omega^2$. We also assume zero volume charge density. Hence, the last term in (2.50) is zero. The wave equation for the electric field becomes

$$\nabla^2\mathbf{E} = (-\mu\epsilon\omega^2 + j\omega\mu\sigma)\mathbf{E} \tag{2.55}$$

Similarly the magnetic field becomes

$$\nabla^2\mathbf{H} = (-\mu\epsilon\omega^2 + j\omega\mu\sigma)\mathbf{H} \tag{2.56}$$

Hence,

$$\nabla^2\mathbf{E} = \gamma^2\mathbf{E}, \quad \nabla^2\mathbf{H} = \gamma^2\mathbf{H} \tag{2.57}$$

where,

$$\gamma^2 = -\mu\epsilon\omega^2 + j\omega\mu\sigma \tag{2.58}$$

If the current density is $\mathbf{J} = \sigma\mathbf{E}$, then for sinusoidal waves

$$\nabla^2\mathbf{J} = \gamma^2\mathbf{J} \tag{2.59}$$

The electric field is also given by

$$\mathbf{E} = -\left(\nabla V + \frac{\partial \mathbf{A}}{\partial t}\right) \tag{2.60}$$

where V is the potential along the conductor.

γ is the *propagation constant* which is complex and may be expressed by

$$\gamma = \alpha + j\beta \tag{2.61}$$

where α = *attenuation constant* and the *phase constant* $\beta = 2\pi/\lambda$, where λ is the wavelength.

$$\gamma^2 = (\alpha + j\beta)^2 = \alpha^2 - \beta^2 + j2\alpha\beta \tag{2.62}$$

Comparing with (2.58)

$$\alpha^2 - \beta^2 = -\mu\epsilon\omega^2\ldots(a), \quad 2\alpha\beta = \omega\mu\sigma\ldots(b) \tag{2.63}$$

From (b), $\beta = \omega\mu\sigma/(2\alpha)$. Substitute into (a) to give

$$\alpha^2 - \frac{(\omega\mu\sigma)^2}{4\alpha^2} = -\mu\epsilon\omega^2, \tag{2.64}$$

Multiplying by α^2 and re-arranging gives

$$\alpha^4 + \mu\epsilon\omega^2\alpha^2 - (\omega\mu\sigma/2)^2 = 0 \tag{2.65}$$

Solving this as a quadratic in α^2, we obtain α and a similar expression for β:

$$\alpha = \omega\sqrt{\mu\epsilon/2}\sqrt{-1 \pm \sqrt{1 + [\sigma/(\omega\epsilon)]^2}} \tag{2.66}$$

$$\beta = \omega\sqrt{\mu\epsilon/2}\sqrt{1 \pm \sqrt{1 + [\sigma/(\omega\epsilon)]^2}} \tag{2.67}$$

The factor $\sigma/(\omega\epsilon)$ is equal to the ratio of the conduction current to the displacement current and is known as the *dissipation factor*. This is discussed in more detail further on in the chapter. If $\sigma = 0$, then $\alpha = 0$ and $\beta = \omega\sqrt{\mu\epsilon}$.

2.4.1 Phase velocity

The general wave (2.51) for sinusoidal waves is

$$\nabla^2 \mathbf{f} = -(\omega/u)^2 \mathbf{f} \tag{2.68}$$

Comparison with (2.55) and (2.57) gives for the phase velocity

$$-(\omega/u)^2 = \gamma^2, \text{ or } u = \frac{j\omega}{\gamma} \tag{2.69}$$

Substituting for γ, then

$$u = \frac{j\omega}{\sqrt{-\omega^2\mu\epsilon + j\omega\mu\sigma}} = \frac{1}{\sqrt{\mu\epsilon}\sqrt{1 - j\sigma/(\omega\epsilon)}} \tag{2.70}$$

If $\sigma = 0$, then $u = 1/\sqrt{\mu\epsilon}$ as for the loss free case.

2.4.2 Refractive index

The refractive index is defined by $n = c/u$, where c is the velocity of light in vacuum. Hence,

$$n^* = -jc\gamma/\omega = -(jc/\omega)\sqrt{-\mu\epsilon\omega^2 + j\omega\mu\sigma} \tag{2.71}$$

Put $n^* = n' - jk''$, where n is the low frequency refractive index and k the absorption coefficient. n and k are known as the *optical constants*. Squaring gives

$$n^{*2} = (n - jk)^2 = n^2 - k^2 - 2jkn \tag{2.72}$$

Squaring (2.71) gives

$$n^{*2} = -(c/\omega)^2(-\mu\epsilon\omega^2 + j\omega\mu\sigma) \tag{2.73}$$

Comparing (2.72) and (2.73) gives

$$n^2 - k^2 = (c/\omega)^2\omega^2\mu\epsilon = \mu_r\epsilon_r \tag{2.74}$$

$$2kn = (c/\omega)^2\omega\mu\sigma = \mu_r\sigma/(\omega\epsilon_o) \tag{2.75}$$

where $c^2 = 1/(\mu_o\epsilon_o)$ has been used. Solving for n and k gives

$$n^2 = \frac{\mu_r\epsilon_r}{2} \pm \frac{1}{2}\sqrt{(\mu_r\epsilon_r)^2 + 4[\sigma\mu_r/(2\omega\epsilon)]^2} \tag{2.76}$$

$$k^2 = -\frac{\mu_r\epsilon_r}{2} \pm \frac{1}{2}\sqrt{(\mu_r\epsilon_r)^2 + 4[\sigma\mu_r/(2\omega\epsilon)]^2} \tag{2.77}$$

If $\sigma = 0$, then $n^2 = \mu_r\epsilon_r$ and $k^2 = 0$. The absorption coefficient is therefore zero, as expected. These latter two equations can be rearranged to give

$$n = \frac{n_o}{\sqrt{2}}\sqrt{1 \pm \sqrt{1 + [\sigma/(\omega\epsilon)]^2}} \tag{2.78}$$

$$k = \frac{n_o}{\sqrt{2}}\sqrt{-1 \pm \sqrt{1 + [\sigma/(\omega\epsilon)]^2}} \tag{2.79}$$

Hence, $n \to n_o$, $k \to 0$ at high frequencies and low σ. At high frequencies ϵ, μ or σ may also be complex. Taking $\epsilon^* = \epsilon' - j\epsilon''$, then

$$n^2 - k^2 = \mu_r\epsilon'_r, \quad 2kn = \mu_r\epsilon''_r + \mu_r\sigma/(\omega\epsilon_o) \tag{2.80}$$

2.4.3   Debye equations

In polar dielectrics, such as liquid water H_2O, the complex permittivity is given by the Debye equations [48]

$$\epsilon' = \frac{\epsilon_s - \epsilon_i}{1 + (\omega\tau)^2} + \epsilon_i, \quad \epsilon'' = \frac{\omega\tau(\epsilon_s - \epsilon_i)}{1 + (\omega\tau)^2} \tag{2.81}$$

where ϵ_s is the static or low frequency permittivity, ϵ_i is the high-frequency permittivity due to induced polarisation and τ is a relaxation time characteristic of the material or liquid. In water at 20°C the approximate values are $\epsilon_s = 80$, $\epsilon_i = 5.0$ and $\tau = 9.2$ ps, [49]. Using these values, the Debye equations for water are shown plotted in Figure 2.4.

Debye plots for water 20°C

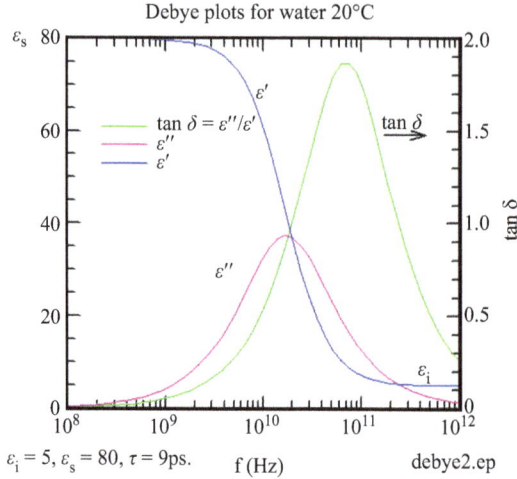

$\varepsilon_i = 5, \varepsilon_s = 80, \tau = 9ps.$ f (Hz)   debye2.ep

Figure 2.4   Debye plot for water

2.4.4   TEM waves

As before consider a transverse electromagnetic (TEM) wave with x-polarised electric vector $\mathbf{E} = E_x\mathbf{a}_x$ and y-polarised magnetic vector $\mathbf{H} = H_y\mathbf{a}_y$. The wave equations are now

$$\frac{\partial^2 E_x}{\partial z^2} = \gamma^2 E_x, \quad \frac{\partial^2 H_y}{\partial z^2} = \gamma^2 H_y \tag{2.82}$$

The solutions are

$$E_x = E_o e^{-\gamma z}, \quad H_y = H_o e^{-\gamma z} \tag{2.83}$$

since $\frac{\partial E_x}{\partial z} = \gamma E_o e^{-\gamma z}$ and $\frac{\partial^2 E_x}{\partial z^2} = \gamma^2 E_o e^{-\gamma z}$, that is, $\frac{\partial^2 E_x}{\partial z^2} = \gamma^2 E_x$ qed.

Substituting for (2.61) gives

$$E_x = E_o e^{-\gamma z} = E_o e^{-(\alpha+j\beta)z} e^{j\omega t} \tag{2.84}$$

This represents a wave travelling in the +z-direction with amplitude which decreases from E_o at the surface ($z = 0$) to $E_o e^{-(\alpha+j\beta)z}$ for $z > 0$. For any given value of $z = z_1$ the waves oscillates about z_1 at an angular frequency ω. This can be plotted in polar form or linear form as shown in Figure 2.5(a) and 2.5(b).

2.4.5   Magnetic and electric fields

For the magnetic field, expanding *curl* **E** in Maxwell's second equation (2.9),

$$\nabla \times \mathbf{E} = \left(\frac{\partial E_z}{\partial y} - \frac{\partial E_y}{\partial z}\right)\mathbf{a}_x + \left(\frac{\partial E_x}{\partial z} - \frac{\partial E_z}{\partial x}\right)\mathbf{a}_y + \left(\frac{\partial E_y}{\partial x} - \frac{\partial E_x}{\partial y}\right)\mathbf{a}_z \tag{2.85}$$

$$= \frac{\partial E_x}{\partial z}\mathbf{a}_y - \frac{\partial E_x}{\partial y}\mathbf{a}_z \tag{2.86}$$

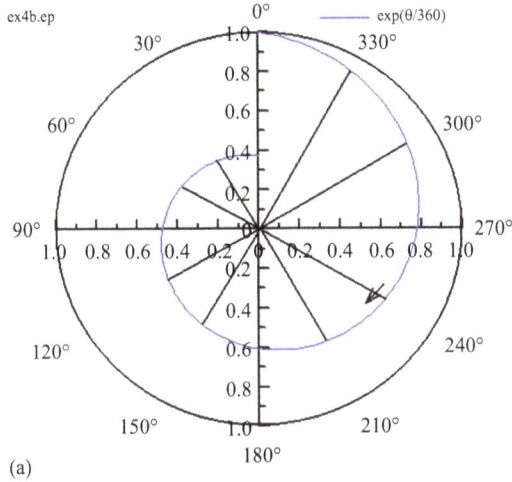

(a)

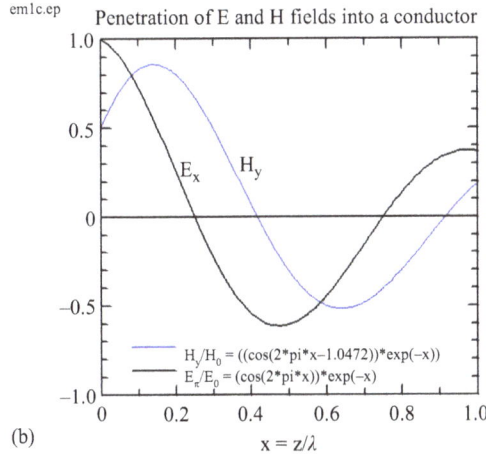

(b)

Figure 2.5   Electric field penetration into a conductor (a) polar plot of amplitude over one cycle. (b) E and H amplitudes over one cycle showing phase delay in H.

Since $E_y = E_z = 0$, and there is no variation of E_x with y, then

$$\frac{\partial E_x}{\partial z}\mathbf{a}_y = -j\omega\mu\mathbf{H} \tag{2.87}$$

$$\mathbf{H} = -\frac{1}{j\omega\mu}\frac{\partial E_x}{\partial z}\mathbf{a}_y \tag{2.88}$$

From (2.83) substitute $\frac{\partial E_x}{\partial z}$ for $-\gamma E_x$ to give

$$\mathbf{H} = \frac{\gamma}{j\omega\mu}E_x\mathbf{a}_y \quad \text{or} \quad H_y = \frac{\gamma}{j\omega\mu}E_x \tag{2.89}$$

2.4.6   Impedance

The impedance of a material is defined by the ratio E/H. Hence, from (2.89)

$$Z = \frac{E_x}{H_y} = \frac{j\omega\mu}{\gamma} \quad \Omega \tag{2.90}$$

Substituting for γ from (2.58) gives

$$Z = \sqrt{\frac{\mu/\epsilon}{1 - j\sigma/(\omega\epsilon)}} \tag{2.91}$$

In terms of the phase angle ϕ between E_x and H_y

$$Z = |Z|\arg\phi = |Z|e^{j\phi} \tag{2.92}$$

where $|Z| = |j\omega\mu/\gamma|$. Hence,

$$H_y = \frac{E_x}{|Z|}e^{-j\phi} \tag{2.93}$$

Substitute for E_x from (2.84) gives

$$H_y = \frac{E_o}{|Z|}e^{-\alpha z}e^{j(\omega t - \beta z - \phi)} \tag{2.94}$$

2.4.7   Summary

For a uniform isotropic lossy material the impedance is everywhere the same but the magnetic field lags behind the electric field. Extraction of the wave constants γ, α, β and Z, is not simple for the general case but fairly straightforward using numerical methods as employed in MATLAB® for example. However, it is important to identify the critical parameters for several areas in electrical and electronic engineering. There are two limiting cases which simplify the analysis: (1) dielectric with loss and (2) conductor with loss.

2.4.8   Dielectric with loss

In this approximation γ in (2.61) is expanded using the binomial expansion where $|x| < 1$

$$(1 - x)^{1/2} \approx 1 - \frac{1}{2}x - \frac{1}{8}x^2 \tag{2.95}$$

$$\gamma = j\omega\sqrt{\mu\epsilon}\sqrt{1 - j\frac{\sigma}{\omega\epsilon}} \tag{2.96}$$

Taking $x = \frac{\sigma}{\omega\epsilon} < 0.1$, then

$$\gamma \approx j\omega\sqrt{\mu\epsilon}\left[1 - j\frac{\sigma}{2\omega\epsilon} + \frac{\sigma^2}{8\omega^2\epsilon^2}\right] = \alpha + j\beta \tag{2.97}$$

Equating real and imaginary parts gives

$$\alpha = \frac{\sigma}{2}\sqrt{\frac{\mu}{\epsilon}} \tag{2.98}$$

$$\beta = \omega\sqrt{\mu\epsilon}\left[1 + \frac{\sigma^2}{8\omega^2\epsilon^2}\right] \approx \omega\sqrt{\mu\epsilon} \tag{2.99}$$

The phase velocity is

$$u = \frac{\omega}{\beta} = \frac{1}{\sqrt{\mu\epsilon}} \tag{2.100}$$

The wave impedance is

$$Z = \sqrt{\frac{\mu/\epsilon}{1 - j\sigma/(\omega\epsilon)}} \tag{2.101}$$

Using the binomial expansion, where $|x| < 1$

$$(1 - x)^{-1/2} \approx 1 + \frac{1}{2}x + \frac{3}{8}x^2 \tag{2.102}$$

$$Z = \sqrt{\mu/\epsilon}\left[1 + j\frac{\sigma}{2\omega\epsilon}\right] \tag{2.103}$$

2.4.9 Conductor with loss

In this case, let $\sigma \gg \omega\epsilon$. Hence, the propagation constant

$$\gamma^2 = j\omega\mu(j\omega\epsilon + \sigma) \tag{2.104}$$

becomes

$$\gamma = \sqrt{j\omega\mu\sigma} \tag{2.105}$$

Using

$$j^{1/2} = \frac{1 + j}{\sqrt{2}} \tag{2.106}$$

Hence,

$$\gamma = \frac{1 + j}{\sqrt{2}}\sqrt{\omega\mu\sigma} \tag{2.107}$$

$$\alpha = \beta = \sqrt{\frac{\omega\mu\sigma}{2}} \tag{2.108}$$

The reciprocal $1/\alpha = \delta$ is the *Skin Depth* with dimensions of metres. This corresponds to the distance the wave propagates before decaying to e^{-1} (about 37 per cent) of its maximum value. This is discussed in more detail further on in the text. From (2.108)

$$\delta = 1/\alpha = \sqrt{\frac{2}{\omega\mu\sigma}} \tag{2.109}$$

The phase velocity is

$$u = \frac{\omega}{\beta} = \sqrt{\frac{2\omega}{\mu\sigma}} \tag{2.110}$$

For $\sigma \gg \omega\epsilon$ the wave impedance, (2.101), becomes

$$Z = \sqrt{\frac{j\omega\mu}{\sigma}} = \sqrt{\omega\mu/(2\sigma)}(1+j) = (1+j)/(\sigma\delta), \quad |Z| = \sqrt{\omega\mu\rho} \tag{2.111}$$

Since $Z = E/H = |Z|arg\,\phi$, then H lags behind E in time by ϕ as shown in Figure 2.5(b). Hence, for a good conductor $tan\,\phi = 1$ and $\phi = 45°$. The phase velocity depends on frequency (2.110), and the wavelength depends on frequency via $\lambda = u/f$. In copper, for example,

$$u = 0.4123f^{1/2}, \quad \lambda = 0.4123f^{-1/2} \tag{2.112}$$

2.4.10   Surface resistivity

For the TEM wave considered previously (2.82) incident on a plane good conductor interface. The average power entering the conductor is [92]

$$P_{av} = \frac{1}{2}Re\int_{S_o+S_z} \mathbf{E} \times \mathbf{H}^*.\mathbf{a}_z ds \tag{2.113}$$

where $\mathbf{E} \times \mathbf{H}^*$ is the Poynting vector, S_o is the wave surface area and S_z the area that the wave enters. For a good conductor the wave fields decrease rapidly as they enter the conductor and S_z can be neglected. Now,

$$\mathbf{E} \times \mathbf{H}^*.\mathbf{a}_z = \mathbf{E}.\mathbf{H}^*\mathbf{a}_z = \mathbf{a}_z \times \mathbf{E}.\mathbf{H}^* \tag{2.114}$$

Also, since the impedance is $Z = \mathbf{E}/\mathbf{H}$, then

$$\mathbf{a}_z \times \mathbf{E}.\mathbf{H}^* = Z\mathbf{H}.\mathbf{H}^* = Z|H|^2 \tag{2.115}$$

Thus,

$$P_{av} = \frac{1}{2}Re\int_{S_o} Z|\mathbf{H}|^2 ds = \frac{R_s}{2}Re\int_{S_o} \mathbf{H}|^2 ds \tag{2.116}$$

where R_s is the surface resistivity of a good conductor.

$$R_s = Re(Z) = Re[(1+j)\sqrt{\frac{\omega\mu}{2\sigma}}] = \sqrt{\frac{\omega\mu}{2\sigma}} = \rho/\delta \;\; \Omega \tag{2.117}$$

that is, the surface resistivity of a good conductor is the ratio of the bulk resistivity to the skin depth: $R_s = \rho/\delta$ Ohms.

Typical values for copper are given in Table 2.1. Equation (2.112) is shown plotted in Figure 2.6.

Generally, in lossy materials, EM waves with different frequencies travel at different phase velocities and are attenuated by different amounts. In an EM pulse, for example, consisting of a spectrum of frequencies, each frequency component will travel at a different velocity, arriving at a distant point at different times. The pulse

Table 2.1 *Phase velocity, wavelength, impedance, skin depth and surface resistivity for copper at 300 K*

f (Hz)	u (m/s)	λ (m)	$|Z|(\Omega)$	δ (m)	$R_s \Omega$
50	2.91	5.83×10^{-2}	2.59×10^{-6}	9.28×10^{-3}	1.83×10^{-6}
10^6	412	4.123×10^{-4}	3.66×10^{-4}	6.56×10^{-5}	2.59×10^{-4}
10^9	1.3×10^4	1.3×10^{-5}	1.16×10^{-2}	2.075×10^{-6}	8.19×10^{-3}

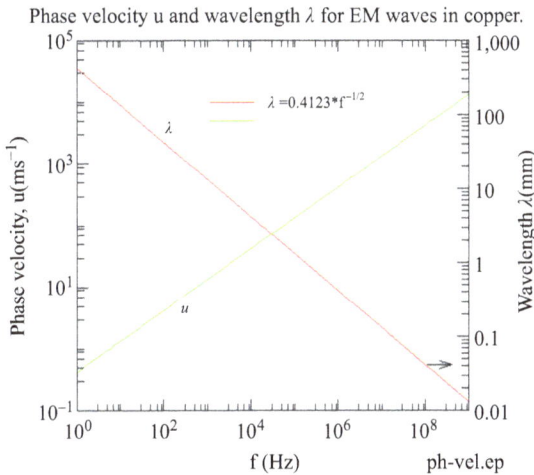

Phase velocity u and wavelength λ for EM waves in copper.

Figure 2.6   Theoretical phase velocity and wavelength for EM waves in copper

shape will therefore change from its original shape. Such a medium is said to be *dispersive*.

2.5   Dissipation factor

The total current J_t is the sum of the ohmic current and the displacement current. The dielectric is lossy with ohmic current loss and loss due to polarisation of the dielectric. In this case, the permittivity is complex $\epsilon = \epsilon' - j\epsilon''$. The total current is given by Maxwell's equation (2.3), which for sine waves is

$$\text{curl } \mathbf{H} = J_t = \sigma E + j\omega(\epsilon' - j\epsilon'')E \tag{2.118}$$

Hence,

$$J_t = (\sigma + \omega\epsilon'')E + j\omega\epsilon' E \tag{2.119}$$

This may be rewritten as

$$J_t = J_c + jJ_d \tag{2.120}$$

where $J_c = (\sigma + \omega\epsilon'')E$ is the conduction current and $J_d = \omega\epsilon'E$ the displacement current. The dissipation factor is defined as the ratio of the conduction current to the displacement current, Figure 2.7. Hence,

$$DF = \frac{J_c}{J_d} = \frac{\sigma + \omega\epsilon''}{\omega\epsilon'} = tan\ \delta \qquad (2.121)$$

The imaginary part of the permittivity (ϵ'') effectively increases the conduction current and loss tangent. Here δ is the phase angle between conduction and displacement currents, not to be confused with the skin depth. A good example of how the dissipation factor varies with frequency is the case of liquid H_2O, Figure 2.8. This result holds up to about 1 GHz. For frequencies up to about 1 MHz, water is a good conductor. For frequencies exceeding 1 MHz, the dissipation decreases, and water becomes a dielectric.

Dissipation factors, also known as loss tangents, are used in the electronics industry to describe the quality of capacitors. Values vary very widely depending on

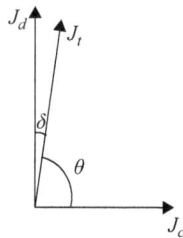

Figure 2.7   Dissipation factor angle δ and power factor angle, θ, defined from the conduction J_c, displacement J_d, and total current J_t

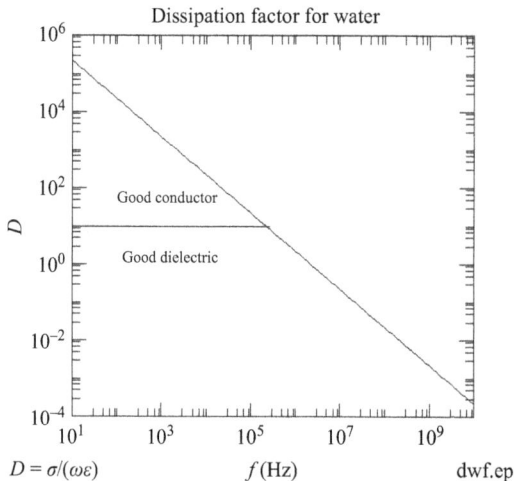

Figure 2.8   Dissipation factor for water at 20°C where $\epsilon_r = \epsilon'_r$, $\epsilon''_r = 0$ for $f \leq 1\ GHz$

the type of capacitor (polarised electrolytic or non-polarised, etc.). *Tan δ* generally increases with frequency and size of capacitor [44]. Capacitors with mica or high-quality plastic film may have a *tan δ* < 10^{-3} or $δ < 0.05°$ measured at 1 kHz. Large value electrolytic capacitors may have *tan δ* > 0.2. Values of *tan δ* are also expressed as a percentage relative to *tan δ* = 1 or $δ = 45°$, that is, a good conductor in (2.121). Some manufacturers quote power factor, which is the cosine of the phase angle $θ$. This assumes the total current $J_t ≈ jωεE$, that is, the J_t vector in Figure 2.7 is nearly vertical for small $δ$. Hence, $cos θ ≈ \frac{σ}{ωε} = tan δ$. In this analysis, no magnetic effects are considered and $μ_r = 1$. Magnetic effects are considered in the following section.

2.6 Quasi-static conditions

The potential drop in a conductor carrying current I is $V_x = IZ$, where Z is the conductor impedance. Taking the case of a good conductor, copper for example, with a TEM wave travelling in the z-direction with amplitude in the x-direction then $V_x = Zσ \int_s E_x \, dydz$ where $Z = (1+j)\sqrt{ωμ/(2σ)}$ and $E_x = E_o exp[(-2π/λ)(1+j)z]$. We assume that the y-dimension is infinite so that E_x is independent of y. Hence,

$$V_x = ZσL_yE_o \int_0^{L_z} e^{(-2π/λ)(1+j)z} \, dz = ZσL_yE_o \frac{1 - e^{(-2π/λ)(1+j)L_z}}{(2π/λ)(1+j)} \qquad (2.122)$$

Substituting for Z gives

$$V_x = \frac{λ}{2π}\sqrt{\frac{ωμ}{2ρ}} L_yE_o[1 - e^{(-2π/λ)(1+j)L_z}] \qquad (2.123)$$

or

$$V_x = L_yE_o[1 - e^{(-2π/λ)(1+j)L_z}] \qquad (2.124)$$

since $λ = 2πδ = 2π\sqrt{\frac{2ρ}{ωμ}}$. This may be put in quadrature form as

$$V_x = L_yE_o\{1 - e^{(-2π/λ)L_z}[cos(2π/λ)L_z - jsin(2π/λ)L_z]\} \qquad (2.125)$$

$$V_x(R) = L_yE_o\{1 - e^{(-2π/λ)L_z}cos(2π/λ)L_z\} \qquad (2.126)$$

$$V_x(I) = L_yE_oe^{(-2π/λ)L_z}sin(2π/λ)L_z \qquad (2.127)$$

Figure 2.9 shows plots of (2.126) and (2.127), normalised to $V_N = V_x/(L_yE_o)$. This shows that for

1. $L_z = λ$ then $V_N(R) = 1, V_N(I) = 0$.
2. $L_z > λ$ then $V_N(R) = 1, V_N(I) = 0$.
3. $L_z ≪ λ$ then $V_N(R) = 0, V_N(I) = 0$.

The quasi-static (qs) approximation assumes that the applied signal wavelength is much greater than the device dimensions. In the above analysis if $λ ≫ L_z$, (case 3) then both $V_N(R)$ and $V_N(I)$ are zero. This implies that the qs condition applies only

Normalised voltage dependence on specimen size

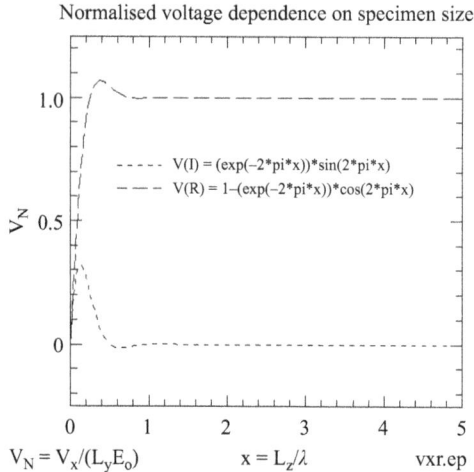

Figure 2.9   Normalised voltage ($V_N = V_x/(L_yE_o)$ quadrature components plotted against ratio of distance in z-direction over wavelength, using (2.126) and (2.127)

to $L_z/\lambda \approx 0$ in Figure 2.9. Taking the first term only of the binomial expansion of (2.124) gives (see proof at the end of this chapter)

$$V_x = L_yE_o\frac{2\pi}{\lambda}(1+j)L_z \tag{2.128}$$

Hence, both the real and imaginary voltages vary linearly with L_z/λ near the origin. Also for a fixed wavelength $V_x \propto L_z$ as expected. We may also substitute $\lambda = 2\pi\delta = 2\pi\sqrt{\frac{2\rho}{\omega\mu}}$ into (2.128) to give

$$V_x = L_yE_oL_z\sqrt{\frac{\omega\mu}{2\rho}} \propto \sqrt{\omega} \tag{2.129}$$

showing that the voltage is proportional to the square root of the frequency. This is also expected since for a good conductor $V = IZ = I(1+j)\sqrt{\omega\mu/(2\sigma)}$, which is proportional to the square root of the frequency.

2.6.1   Equivalent circuit approximations-lossy conductor and superconductor

For *quasi-static conditions*, that is, wavelengths much larger than the device dimensions, the E and H fields are similar to static fields [43], and unique I and V values can be identified. Figure 2.10 shows the equivalent circuit of a device or material. R_n is the normal resistance due to ohmic conduction, L_n the associated normal inductance due to magnetising current, and C_n the normal capacitance due to displacement current. In this model, both the permeability and the permittivity are assumed to be complex. L_λ is the ideal superinductance (see Superconductivity in 3.3.2). Also included is

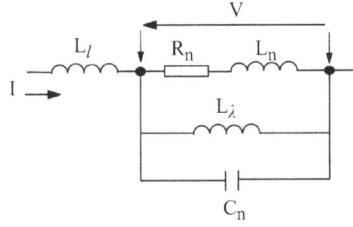

Figure 2.10   Equivalent circuit of device showing lead inductance L_l, normal inductance, resistance and capacitance L_n, R_n, C_n, respectively, and superinductance L_λ

the total inductance due to leads, although the four-terminal technique tends to minimise this inductance. The total current driven through the device is given by circuit theory as

$$I = \frac{V}{Z_{L_n}} + j\omega C_n V + \frac{V}{j\omega L_\lambda} = YV \qquad (2.130)$$

where Y is the admittance of the device and

$$Z_{L_n} = R_n + j\omega L_n, \quad Z_{L_n}^2 = R_n^2 + (\omega L_n)^2 \qquad (2.131)$$

Rationalising and re-arranging gives

$$Y = \frac{R_n}{Z_{L_n}^2} + j\omega \left[C_n - \frac{L_n}{Z_{L_n}^2} - \frac{1}{\omega^2 L_\lambda} \right] \qquad (2.132)$$

$$\mathbf{J} = [\sigma_e + j\omega\epsilon - j/(\omega L_e)]\mathbf{E} \qquad (2.133)$$

where

$$\sigma_e = \frac{R_n}{Z_{L_n}^2}g, \quad \frac{1}{L_e} = \left[\frac{\omega^2 L_n}{Z_{L_n}^2} + \frac{1}{L_\lambda} \right]g \qquad (2.134)$$

$$L_\lambda = \mu\lambda^2 g, \quad L_n = \mu a^2 g \qquad (2.135)$$

where λ is the superconducting penetration depth, the equation for L_n is for a slab of material where g is a geometrical factor $g = d/A$. d is the distance between voltage contacts, a is the half-thickness of the slab, and A is the cross-sectional area perpendicular to the current path $(A = 2aX)$, Figure 2.11. A more accurate equation for the low frequency inductance of a rectangular slab is [46] originally given by Terman [45], here converted to SI units,

$$L = 2 \left[\ln\frac{l}{w+t} + 1.19 + 0.22\frac{w+t}{l} \right] \quad nH \ cm^{-1} \qquad (2.136)$$

where l, w, t are the length, width and thickness of the slab, respectively. For a square sheet $l = w \gg t$ then $L_n = 2.82$ nH cm^{-1}. These values decrease by approximately 10 per cent at microwave frequencies [46].

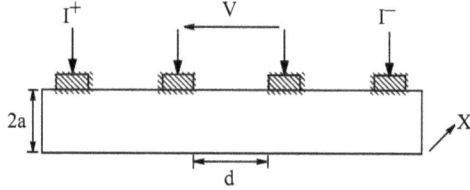

Figure 2.11   *Rectangular slab with four contact pads for the measurement of impedance. The current I is injected into contact I^+ and extracted from contact I^-. The potential drop V is measured between the inner voltage contacts using a high input impedance voltmeter. This minimises contact resistance effects.*

2.6.2   Impedance

Re-write (2.133) as

$$\mathbf{J} = j\omega[\sigma_e/(j\omega) + \epsilon - 1/(\omega^2 L_e)]\mathbf{E} = j\omega\epsilon^*\mathbf{E} \tag{2.137}$$

where the effective permittivity ϵ^* is

$$\epsilon^* = \epsilon - 1/(\omega^2 L_e) - j\sigma_e/(\omega). \tag{2.138}$$

The impedance for the loss free case is from (2.46)

$$Z = \frac{E_x}{H_y} = \sqrt{\frac{\mu}{\epsilon}} \tag{2.139}$$

Substituting for ϵ^* gives

$$Z = \sqrt{\frac{\mu}{\epsilon^*}} = \sqrt{\mu/[\epsilon - 1/(\omega^2 L_e) - j\sigma_e/(\omega)]} = \frac{j\omega\mu}{\gamma} \tag{2.140}$$

where the propagation constant is

$$\gamma = \sqrt{-\omega^2\mu\epsilon + \mu/L_e + j\omega\mu\sigma_e} \tag{2.141}$$

If there is no inductive current, that is, $\mu/L_e = 0$ then

$$\gamma = \sqrt{-\omega^2\mu\epsilon + j\omega\mu\sigma_e} \tag{2.142}$$

which agrees with the lossy medium case equation (2.58).

2.6.3   Complex μ and ϵ

In this section, the material is considered to have complex permeability and permittivity. The total current is now

$$I = \frac{V}{Z_{L_n}^*} + j\omega C_n^* V + \frac{V}{j\omega L_\lambda^*} = Y^* V \tag{2.143}$$

where the stars * indicate complex components. The normal inductive impedance becomes

$$Z_{Ln} = R_n + j\omega L_n^* = R_n + j\omega(L_n' - jL_n'') \tag{2.144}$$

Hence,

$$Z_{Ln}^* = R_n + \omega L_n'' + j\omega L_n' \tag{2.145}$$

where

$$L_n^* = L_n' - jL_n'', \quad L_n' = \mu'a^2g, \quad L_n'' = \mu''a^2g \tag{2.146}$$

The superinductance is

$$L_\lambda^* = L_\lambda' - jL_\lambda'' = \mu'\lambda^2g - j\mu''\lambda^2g \tag{2.147}$$

The current density becomes

$$\mathbf{J}^* = [\sigma_e^* + j\omega\epsilon' - j/(\omega L_e^*)]\mathbf{E} \tag{2.148}$$

where

$$\sigma_e^* = \frac{R_n g}{Z_{Ln}^{*2}} + \frac{L_\lambda'' g}{\omega L_\lambda^{*2}} + \omega\epsilon'' \tag{2.149}$$

$$\frac{1}{L_e^*} = \frac{\omega^2 L_n' g}{Z_{Ln}^{*2}} + \frac{L_\lambda' g}{L_\lambda^{*2}} \tag{2.150}$$

2.6.4 Mutual coupling effect

The impedance of two inductors in parallel is given by Raven [47]

$$Z = \frac{Z_1 Z_2 - X_M^2}{Z_1 + Z_2 - X_M^2} \tag{2.151}$$

where $X_M = j\omega M$ is the mutual inductive reactance coupling the two inductors. For the case of normal current coupling with a supercurrent, Figure 2.10, then

$$Z_1 = Z_n = R_n + j\omega L_n, \quad Z_2 = Z_\lambda = R_\lambda + j\omega L_\lambda, \quad X_M = j\omega M \tag{2.152}$$

where Z_n, R_n, L_n are normal impedance, resistance and inductance, respectively. Z_λ, R_λ, L_λ are superconducting impedance, resistance and inductance respectively. Any or all of these parameters may be complex due to complex μ and ϵ. However, the normal conductivity σ_n is only likely to be complex at optical frequencies, far exceeding this quasi-static approximation (see page 40). Although ideally R_λ is zero, in practice, some loss may occur due to vortex dissipation.

The mutual coupling is defined by

$$M = k\sqrt{L_1 L_2} = k\sqrt{L_n L_\lambda} \tag{2.153}$$

where k is the coupling coefficient, which is expected to be close to unity for a simple superconductor. The total impedance, including the parallel leakage impedance Z_l is

$$Z_T = \frac{ZZ_l}{Z + Z_l} = \frac{Z}{Z/Z_l + 1} \tag{2.154}$$

The admittance $Y = 1/Z$ is

$$Y_T = Z^{-1} + Z_l^{-1} \tag{2.155}$$

For a simple rectangular device, Figure 2.11, length d and current path crossectional area A; the specific admittance is $y_T = Y_T d/A = Y_T g$ where $g = d/A$. The current density is then

$$\mathbf{J} = y_T \mathbf{E} = (y + y_l)\mathbf{E} = Y_T g \mathbf{E} \tag{2.156}$$

To check this put $X_M = 0, R_\lambda = 0$ and $y_l = j\omega C_n g = j\omega\epsilon$. Hence,

$$\mathbf{J} = \left[\frac{R_n}{Z_{L_n}^2} + j\omega C_n - j\left(\frac{\omega L_n}{Z_{L_n}^2} + \frac{1}{\omega L_\lambda} \right) \right] g\mathbf{E} = Y_T g\mathbf{E} \tag{2.157}$$

$$\mathbf{J} = [\sigma_e + j\omega\epsilon - j/(\omega L_e)]\mathbf{E} = y_T \mathbf{E} \tag{2.158}$$

where $\sigma_e = R_n g/Z_{L_n}^2$, $\epsilon = C_n g$ and $1/L_e = [\omega^2 L_n/Z_{L_n}^2 + 1/L_\lambda]g$. This agrees with (2.133) and (2.134). Note that if ϵ or μ are complex then the real and imaginary parts of (2.157) and (2.158) are modified.

2.6.5 Normal state with complex μ and ϵ

In this case assume $\lambda \to \infty$. Hence $L_\lambda = 0$ in (2.135). Introduction of complex μ gives

$$L_n = \mu a^2 g = (\mu' - j\mu'')L_{n1}   L_{n1} = \mu_o a^2 g \tag{2.159}$$

$g = d/A$, A is the crossectional area of the rectangular device length d, Figure 2.11.

$$\sigma_e = \frac{R_n}{R_n^2 + [\omega L_{n1}(\mu' - j\mu'')]^2} = \frac{R_n}{A1 - jB} \tag{2.160}$$

$$\frac{1}{L_e} = \frac{\omega^2 L_n g}{R_n^2 + (\omega L_n)^2} = \frac{\omega^2 L_n g}{A1 - jB} \tag{2.161}$$

$$\mathbf{J} = \left[R_n g \frac{A1 + jB}{A1^2 + B^2} + j\omega(\epsilon' - j\epsilon'') - j\omega L_n g \frac{A1 + jB}{A1^2 + B^2} \right] \mathbf{E} \tag{2.162}$$

Hence,

$$\mathbf{J} = (CA1 + DB + \omega\epsilon'')\mathbf{E} + j(BC - DA1 + \omega\epsilon')\mathbf{E} \tag{2.163}$$

$$\tan\delta = \frac{(CA1 + DB + \omega\epsilon'')\mathbf{E}}{BC - DA1 + \omega\epsilon'} \tag{2.164}$$

where

$$A1 = R_n^2 + (\omega L_{n1})^2(\mu'^2 + \mu''^2), \quad B = 2\mu'\mu''(\omega L_{n1})^2 \tag{2.165}$$

$$C = \frac{R_n g}{A1^2 + B^2}, \quad D = \frac{\omega L_n g}{A1^2 + B^2} \tag{2.166}$$

2.7   Proof of equation (2.128)

Re-writing (2.124) as

$$V_x = L_y E_o [1 - e^x] \tag{2.167}$$

where

$$x = (-2\pi/\lambda)(1 + j)L_z \tag{2.168}$$

The binomial expansion of e^x is

$$e^x = 1 + x/1! + x^2/2! + x^3/3! + \cdots \tag{2.169}$$

The first term of this expansion leads to

$$V_x = L_y E_o [1 - e^x] = L_y E_o [1 - (1 + x)] = L_y E_o - L_y E_o (1 + x) \tag{2.170}$$

Substituting for (2.168) gives (2.128)

$$V_x = -L_y E_o [(-2\pi/\lambda)(1 + j)]L_z = L_y E_o \frac{2\pi}{\lambda}(1 + j)L_z \tag{2.171}$$

Chapter 3
Power dissipation and Poynting's theorem

3.1 Steady state dc power dissipation

The work done per second or energy per second, that is, power dissipated is

$$p = dW/dt = \mathbf{F}.d\mathbf{r}/dt = \rho_v \mathbf{E}.\mathbf{v} = \mathbf{E}.\mathbf{J} \tag{3.1}$$

where $\mathbf{r}$ is the distance moved by the charge, $\mathbf{v}$ its velocity and $\mathbf{J}$ the electric current density $\mathbf{J} = \rho_v \mathbf{v}$. Hence, the power dissipated per unit volume is

$$p = \mathbf{E}.\mathbf{J}, \quad \text{watts } \text{m}^{-3} \tag{3.2}$$

This energy loss per second can then be equated to the energy or power in the electric and magnetic fields, leading to Poynting's theorem.

For uniform current flow $J = I/A$, where A is the cross-sectional area of the conductor length l, $\mathbf{E} = -\nabla\psi = V/l$ volts per m. Hence, the power dissipated in volume lA is

$$p = \mathbf{E}.\mathbf{J}\, lA = (V/l)(I/A)lA = VI, \quad \text{watts} \tag{3.3}$$

In an alternative approach, the work done on moving a charge q through a potential difference V is $W = qV$ joules or eV electron volts. The current is $I = q/t$. Hence, the power dissipated is

$$p = W/t = VI, \quad \text{watts} \tag{3.4}$$

The first approach is important because it can immediately be related to Maxwell's equations as shown in the section on Poynting's theorem 3.3.

3.2 Power dissipation in the time and frequency domain

3.2.1 Time domain

Time-dependent electric and magnetic fields are common and range from transient effects such as switching on/off lights or motors, charging and discharging capacitors, and transient phenomena, including electrical breakdown in insulators and air, such as lightning. In general, power dissipated or delivered to a load is determined from the product of the time-dependent functions $v(t)$ and $i(t)$. The energy delivered to a load for $t_1 < t < t_2$ is then

$$w = \int_{t_1}^{t_2} v(t)i(t)dt \tag{3.5}$$

3.2.2   Frequency domain

For sinusoidal currents and voltages in the frequency domain, consider the case of a resistor in series with an inductor with sinusoidal current i and sinusoidal voltage v which leads the current by a phase angle ϕ, [27]. The maximum values are i_m and v_m, and the instantaneous values are

$$i = i_m \sin\theta, \quad v = v_m \sin(\theta + \phi) \tag{3.6}$$

where $\theta = \omega t$. The instantaneous power is

$$p = vi = v_m i_m \sin\theta \sin(\theta + \phi) \tag{3.7}$$

$$p = vi = v_m i_m \sin\theta[\sin\theta \cos\phi + \cos\theta \sin\phi] \tag{3.8}$$

Substituting

$$\sin^2\theta = \frac{1}{2}(1 - \cos 2\theta), \quad \sin\theta \cos\theta = \frac{1}{2}\sin 2\theta \tag{3.9}$$

gives

$$p = v_m i_m \left[\frac{1}{2}(1 - \cos 2\theta)\cos\phi + \frac{1}{2}\sin 2\theta \sin\phi\right] \tag{3.10}$$

$$p = (v_m i_m/2)[\cos\phi - \cos 2\theta \cos\phi) + \sin 2\theta \sin\phi] \tag{3.11}$$

Now the *rms* current and voltages are given by

$$I = I_{rms} = i_m/\sqrt{2}, \quad V = V_{rms} = v_m/\sqrt{2}, \quad IV = V_{rms}I_{rms} = v_m i_m/2 \tag{3.12}$$

Hence,

$$p = VI\cos\phi - VI\cos(2\theta - \phi) \tag{3.13}$$

The *power factor* is defined as

$$PF = \frac{mean\ power}{rms\ volt\ amps} = \frac{P}{VI} \tag{3.14}$$

For the RL circuit

$$PF = cos\phi \tag{3.15}$$

The mean power is

$$P = VI\cos\phi \tag{3.16}$$

3.2.3   Instantaneous power

Expanding (3.13) gives

$$p = VI\cos\phi - VI\cos\phi\cos 2\theta + VI\sin\phi\sin 2\theta \tag{3.17}$$

Thus

$$p = \textit{steady power in } R - \textit{osc. power in } R \textit{ at } 2f + \textit{osc. power in } X \textit{ at } 2f \quad (3.18)$$

The terms in (3.17) are:

1. *First term.* Steady power dissipated in the resistive or Ohmic elements

$$P = I^2R = VI\cos\phi \quad \text{watts} \tag{3.19}$$

2. *Second term.* Oscillatory power at $2f$ dissipated in R. Does not involve stored energy and is usually ignored.
3. *Third term.* Oscillatory power at $2f$ dissipated in reactive elements X. Important in power systems and associated with reactive stored energy in L or C. It has a maximum value

$$Q = VI\sin\phi = I^2X \quad \text{vars} \tag{3.20}$$

The instantaneous power is then

$$p = P + Q \tag{3.21}$$

where P and Q are given by (3.19) and (3.20), respectively. These may be expressed in phasor form

$$\mathbf{S} = P + jQ = VI(\cos\phi + j\sin\phi) \tag{3.22}$$

But note that

1. P is the steady state non-oscillatory power measured in watts.
2. Q is the maximum value of double frequency oscillatory power flow into the reactance X measured in vars.

None of these P, Q or S are sine components at frequency f with instantaneous time value.

Now let $\mathbf{I}$ have a phase angle ϕ ahead of $\mathbf{V}$ which implies that the circuit is predominantly capacitive. Hence,

$$\phi = \phi_I - \phi_V \tag{3.23}$$

Let the current and voltage phasors be given by

$$\mathbf{I} = a + jb, \quad \mathbf{V} = c + jd \tag{3.24}$$

$$P = VI\cos\phi = VI\cos(\phi_I - \phi_V) \tag{3.25}$$

$$= VI[\cos\phi_I\cos\phi_V + \sin\phi_I\sin\phi_V] \tag{3.26}$$

$$= VI[\frac{a}{I}\cdot\frac{c}{V} + \frac{b}{I}\cdot\frac{d}{V}] = (ac + bd), \quad \text{watts} \tag{3.27}$$

$$Q = VI\sin\phi = VI\sin(\phi_I - \phi_V) \tag{3.28}$$

$$= VI[\sin\phi_I\cos\phi_V - \cos\phi_I\sin\phi_V] \tag{3.29}$$

$$= VI[\frac{b}{I}\cdot\frac{c}{V} - \frac{a}{I}\cdot\frac{d}{V}] = (bc - ad), \quad \text{vars} \tag{3.30}$$

The phasor product $\mathbf{I}.\mathbf{V}$ should give dimensions of power. Thus

$$\mathbf{I}.\mathbf{V} = (a + jb)(c + jd) = (ac - bd) + j(bc + ad) \neq P + jQ \tag{3.31}$$

Hence the real part of $\mathbf{I}.\mathbf{V}$ does not give P and the imaginary part does not give Q. To overcome this problem we take the complex conjugate of either $\mathbf{I}$ or $\mathbf{V}$ but not both. The two schemes are

1. $\mathbf{S} = \mathbf{V}^*.\mathbf{I} = (c - jd)(a + jb) = (ac + bd) + j(bc - ad)$
   $= P + jQ, \quad$ *as equation* (3.22). $\tag{3.32}$

2. $\mathbf{S} = \mathbf{V}.\mathbf{I}^* = (c + jd)(a - jb) = (ac + bd) + j(ad - bc)$
   $= P - jQ, \quad$ *different.* $\tag{3.33}$

In scheme (1) using the supply voltage phasor as the reference (datum) then if

$$\mathbf{V} = V + j0 = V - j0 = \mathbf{V}^* \tag{3.34}$$

$$\mathbf{I} = I_P \pm jI_Q = I(\cos\phi \pm j\sin\phi) \tag{3.35}$$

$$\mathbf{S} = V^*\mathbf{I} = V\mathbf{I} = VI_P \pm jVI_Q = P \pm jQ \tag{3.36}$$

Hence, the P and Q components of $\mathbf{S}$ are directly related to the current components I_P and I_Q, their magnitude and sign.

In Scheme (2), Q is positive for inductive loads and lagging current. Since a very large proportion of industrial loads are, inductive then with $+Q$ the vars are positive. If Scheme (1) was employed, it would be necessary to nearly always refer to negative vars.

3.3   Power Flow-Poynting's theorem

This section considers power flow in electromagnetic fields [57]. The theory considers the additional energy due to electromagnetic effects. It does not include other sources of energy dissipation due to mechanical and acoustic energy inputs, [58]. Consider an electrical power source p_s applying energy to a material with volume V. The energy dissipated W per unit volume is the work done by the applied field $\mathbf{E}$ in moving charge a distance $\mathbf{d}$

$$W = \rho_v \mathbf{E}.\mathbf{d} \tag{3.37}$$

where ρ_v is the volume charge density. The rate of energy dissipated is the power dissipated

$$\frac{dW}{dt} = p_d = \rho_v \mathbf{E}.\frac{d\mathbf{d}}{dt} = \rho_v \mathbf{E}.\mathbf{v} \tag{3.38}$$

where $\mathbf{v}$ is the charge velocity. Also by definition the current density is $\mathbf{J} = \rho_v \mathbf{v}$. Hence, the power dissipated per unit volume is

$$p_d = \mathbf{E}.\mathbf{J} \tag{3.39}$$

From Maxwell's Ampere equation $\mathbf{J} = \nabla \times \mathbf{H} - \dot{\mathbf{D}}$ where $\dot{\mathbf{D}} = \partial\mathbf{D}/\partial t$,

$$p_d = \mathbf{E}.(\nabla \times \mathbf{H} - \dot{\mathbf{D}}). \tag{3.40}$$

Using the vector identity

$$\nabla.(\mathbf{E} \times \mathbf{H}) = \mathbf{H}.(\nabla \times \mathbf{E}) - \mathbf{E}.(\nabla \times \mathbf{H}) \tag{3.41}$$

$$p_d = [\mathbf{H}.(\nabla \times \mathbf{E}) - \nabla.(\mathbf{E} \times \mathbf{H}) - \mathbf{E}.\dot{\mathbf{D}}] \tag{3.42}$$

Substituting for Maxwell's equation $\nabla \times \mathbf{E} = -\dot{\mathbf{B}}$ gives

$$p_d = -[\mathbf{H}.\dot{\mathbf{B}} + \mathbf{E}.\dot{\mathbf{D}} + \nabla.(\mathbf{E} \times \mathbf{H})] \tag{3.43}$$

Integrating equation (3.43) over volume V gives

$$P_d = -[\int_V [\mathbf{H}.\dot{\mathbf{B}} + \mathbf{E}.\dot{\mathbf{D}}]dV + \int_V \nabla.(\mathbf{E} \times \mathbf{H})dV] \tag{3.44}$$

Using Gauss's theorem

$$\int_V \nabla.(\mathbf{E} \times \mathbf{H})dV = \oint_S (\mathbf{E} \times \mathbf{H}).d\mathbf{s} \tag{3.45}$$

gives

$$P_d = -[\int_V [\mathbf{H}.\dot{\mathbf{B}} + \mathbf{E}.\dot{\mathbf{D}}]dV + \oint_S (\mathbf{E} \times \mathbf{H}).d\mathbf{s}] \tag{3.46}$$

Integrating (3.39) over volume V gives

$$P_d = \int_V \mathbf{E}.\mathbf{J}dV \tag{3.47}$$

Equating (3.46) and (3.47) give

$$\int_V \mathbf{E}.\mathbf{J}dV = -[\int_V (\mathbf{H}.\dot{\mathbf{B}} + \mathbf{E}.\dot{\mathbf{D}})dV + \oint_S (\mathbf{E} \times \mathbf{H}).d\mathbf{s}] \tag{3.48}$$

Thus, the total power dissipated is equal to the sum of the rate of decrease of the electric and magnetic energies in the volume V and the power flow out across the surface S. From this (3.48), the flow out can be expressed by

$$\oint_S (\mathbf{E} \times \mathbf{H}).d\mathbf{s} = -[\int_V (\mathbf{H}.\dot{\mathbf{B}} + \mathbf{E}.\dot{\mathbf{D}} + \mathbf{E}.\mathbf{J})dV] \tag{3.49}$$

This shows that the total power flow out across the closed surface S is equal to the rate of decrease of the energy in the electric and magnetic fields plus the Joule power dissipation in the volume V. This is *Poynting's Theorem*. The vector

$$\mathbf{S} = \mathbf{E} \times \mathbf{H} \tag{3.50}$$

is Poynting's vector. This has dimensions of *Watts*/m^2 and its direction is that of power flow. The symbol $\mathbf{S}$ is used for the Poynting vector not to be confused with the area vector $\mathbf{S}$. For this reason some authors use the symbol $\mathcal{P}$.

For the case of energy flowing into a material (3.49) can be rewritten as

$$s = -\oint_S (\mathbf{E} \times \mathbf{H}).d\mathbf{s} = \int_V (\mathbf{H}.\dot{\mathbf{B}} + \mathbf{E}.\dot{\mathbf{D}} + \mathbf{E}.\mathbf{J})dV \tag{3.51}$$

The first equation on the left is the power into the material-negative IN, positive OUT. The first term on the right is ohmic loss and the second term on the right is the EM power.

In applying Poynting's theorem **E** and **H** must be causally related i.e both must arise from the same EM source. Electric current (ac or dc) flowing towards a load in a wire pair will have a magnetic field surrounding the wires coupled to an electric field between the forward and return wires. This will give rise to a Poynting Vector with direction towards the load and magnitude proportional to the power dissipated in the load. A capacitor containing a steady charge Q and zero current flow, placed in an external magnetic field, is a case where the Poynting vector is zero.

3.3.1   Poynting's theorem – alternative derivation

The rate of flow of electromagnetic energy density (power) outwards from a volume V is equal to the rate of decrease of the total energy

$$p = -\left[\frac{\epsilon}{2}\frac{\partial E^2}{\partial t} + \frac{\mu}{2}\frac{\partial H^2}{\partial t} + \mathbf{E.J}\right] \tag{3.52}$$

where the first term is due to the energy density in the E and H fields and the second term is Joule heating due to the current flow. Since $ydy = dy^2/2$ then

$$p = -\left[\epsilon E.\frac{\partial E}{\partial t} + \mu H.\frac{\partial H}{\partial t} + \mathbf{E.J}\right] \tag{3.53}$$

$$= -\left[E.\frac{\partial D}{\partial t} + H.\frac{\partial B}{\partial t} + \mathbf{E.J}\right] \tag{3.54}$$

$$= -\left[E.(\frac{\partial D}{\partial t} + J) + H.\frac{\partial B}{\partial t}\right] \tag{3.55}$$

Using Maxwell's equations $\mathbf{J} = \nabla \times \mathbf{H} - \dot{\mathbf{D}}$ and $\nabla \times \mathbf{E} = -\dot{\mathbf{B}}$. This gives

$$p = -[\mathbf{E}.(\dot{\mathbf{D}} + \nabla \times \mathbf{H} - \dot{\mathbf{D}}) - \mathbf{H}.(\nabla \times \mathbf{E}] \tag{3.56}$$

Hence,

$$p = -[\mathbf{E}.(\nabla \times \mathbf{H}) - \mathbf{H}.(\nabla \times \mathbf{E})] \tag{3.57}$$

Using a vector identity the latter equation becomes

$$p = \nabla.(\mathbf{E} \times \mathbf{H}) \tag{3.58}$$

Integrating (3.58) over the volume enclosing the energy and using the divergence theorem gives

$$P = \oint_S (\mathbf{E} \times \mathbf{H}).d\mathbf{s} \tag{3.59}$$

This is positive and equals the total flow of power out over the surface S of the volume V. Integrating (3.54) over the same volume V gives

$$P = -\int_V [\mathbf{E}.\dot{\mathbf{D}} + \mathbf{H}.\dot{\mathbf{B}} + \mathbf{E.J}]dV \tag{3.60}$$

This is the same power but it is negative and represents the decrease in power contained in the volume *V*. Equating (3.59) and (3.60) gives

$$\oint_S (\mathbf{E} \times \mathbf{H}).d\mathbf{s} = -[\int_V (\mathbf{H}.\dot{\mathbf{B}} + \mathbf{E}.\dot{\mathbf{D}} + \mathbf{E}.\mathbf{J})dV]$$ (3.61)

which agrees with (3.49). This alternative approach assumes equations for the energy density in the *E* and *H* fields and the Joule equation E.J. The previous approach, [59], only assumed E.J as a dissipation source and derives Poynting's theorem from this.

3.3.2 Superconductivity

. The most widely used theories to describe superconductivity up to about 1986 were the thermodynamic theory of Gorter and Casimir [52], the electromagnetic theory of London and London [51], the theory based on second-order phase transitions by Ginzburg and Landau [53], and superconductivity at a microscopic level due to Bardeen *et al.* [54]. A general theory for the high-temperature superconductors discovered in 1986 [55] remains to be fully developed.

In a superconductor which obeys London theory [51], the current density is given by

$$\mathbf{J_s} = -qn_s\mathbf{v_s}$$ (3.62)

where n_s and $\mathbf{v_s}$ are the supercarrier density and velocity respectively. Assuming no collisions the equation of motion is

$$m^* \frac{d\mathbf{v_s}}{dt} = -q\mathbf{E}$$ (3.63)

Hence,

$$\frac{d\mathbf{J_s}}{dt} = -qn_s \frac{d\mathbf{v_s}}{dt} = \frac{q^2 n_s}{m^*}\mathbf{E}$$ (3.64)

This leads to London's First Equation

$$\frac{d\mathbf{J}_s}{dt} = \frac{\mathbf{E}}{\Lambda}$$ (3.65)

where $\Lambda = m^*/(n_s q^2) = \mu\lambda_L^2$ and λ_L is the London penetration depth. The power dissipated is then

$$p_d = \mathbf{E}.\mathbf{J}_n + \mathbf{E}.\mathbf{J}_s = \mathbf{E}.\mathbf{J}_n + \Lambda\mathbf{J}_s\frac{d\mathbf{J}_s}{dt}$$ (3.66)

$$p_d = \mathbf{E}.\mathbf{J}_n + \mathbf{E}.\mathbf{J}_s = \mathbf{E}.\mathbf{J}_n + \frac{\Lambda}{2}\frac{d\mathbf{J}_s^2}{dt}$$ (3.67)

The first term is due to dissipation of the normal fluid. The second term is due to dissipation of the superfluid. This is time or frequency dependent and is zero in the steady state. Poynting's equation (3.51) then becomes

$$s = \int_V [\mathbf{E}.\mathbf{J}_n + \Lambda\mathbf{J}_s.\dot{\mathbf{J}}_s + \mathbf{H}.\dot{\mathbf{B}} + \mathbf{E}.\dot{\mathbf{D}}]dV$$ (3.68)

For linear Ohmic materials this can be written as

$$s = \int_V \frac{J_n^2}{\sigma_n} dV + \frac{1}{2}\frac{d}{dt}\int_V (\epsilon E^2 + \mu H^2 + \Lambda J_s^2) dV \tag{3.69}$$

Thus, the supercurrent only contributes to the time dependent term or inertial term of the power flow.

3.3.3   Complex Poynting vector

In this section we consider power flow in lossy materials where the applied field varies sinusoidally with time defined by

$$\mathbf{E} = \mathbf{E}_o \cos(\omega t + \theta), \quad \mathbf{H} = \mathbf{H}_o \cos(\omega t + \phi). \tag{3.70}$$

or

$$\mathbf{E} = \mathbf{E}_o \mathcal{R}_e(e^{j(\omega t + \theta)}), \quad \mathbf{H} = \mathbf{H}_o \mathcal{R}_e(e^{j(\omega t + \phi)}). \tag{3.71}$$

The power flow is given by the real part of the Poynting vector

$$\mathbf{S} = \mathcal{R}_e(\mathbf{E}) \times \mathcal{R}_e(\mathbf{H}) = (\mathbf{E}_o \times \mathbf{H}_o)\cos(\omega t + \theta)\cos(\omega t + \phi) \tag{3.72}$$

$$= \frac{1}{2}(\mathbf{E}_o \times \mathbf{H}_o)[\cos(2\omega t + \theta + \phi) + \cos(\theta - \phi)] \tag{3.73}$$

The time average Poynting vector is

$$<\mathbf{S}> = \frac{1}{T}\int_0^T \mathbf{S} dt = \frac{1}{2}(\mathbf{E}_o \times \mathbf{H}_o)\cos(\theta - \phi) \tag{3.74}$$

since the time average of the first cosine term is zero. Put in terms of the phasors

$$\hat{\mathbf{E}} = \mathbf{E}_o e^{j\theta}, \quad \hat{\mathbf{H}} = \mathbf{H}_o e^{j\phi} \tag{3.75}$$

Hence,

$$<\mathbf{S}> = \mathcal{R}_e \frac{1}{2}[\hat{\mathbf{E}} \times \hat{\mathbf{H}}^*], \quad <\mathbf{S}> = \mathcal{R}_e \frac{1}{2}[\hat{\mathbf{E}}^* \times \hat{\mathbf{H}}] \tag{3.76}$$

where * is the complex conjugate. Assuming a volume of material V bounded by surface S, the total power flow out of S is

$$\oint_S <\mathbf{S}> .d\mathbf{s} = \oint_S \frac{1}{2}[\hat{\mathbf{E}} \times \hat{\mathbf{H}}^*].d\mathbf{s} = \frac{1}{2}\int_V \nabla.(\hat{\mathbf{E}} \times \hat{\mathbf{H}}^*) dV \tag{3.77}$$

Using the vector identity

$$\nabla.(\hat{\mathbf{E}} \times \hat{\mathbf{H}}^*) = \hat{\mathbf{H}}^*.(\nabla \times \hat{\mathbf{E}}) - \hat{\mathbf{E}}.(\nabla \times \hat{\mathbf{H}}^*) \tag{3.78}$$

and substituting for Maxwell's equations

$$\nabla \times \hat{\mathbf{E}} = -j\omega\mu\hat{\mathbf{H}}, \quad \nabla \times \hat{\mathbf{H}}^* = \hat{\mathbf{J}}^* - j\omega\epsilon^*\hat{\mathbf{E}}^* \tag{3.79}$$

gives

$$\nabla.(\hat{\mathbf{E}} \times \hat{\mathbf{H}}^*) = -\hat{\mathbf{E}}.\hat{\mathbf{J}}^* + j\omega\epsilon^*\hat{\mathbf{E}}.\hat{\mathbf{E}}^* - j\omega\mu\hat{\mathbf{H}}.\hat{\mathbf{H}}^* \tag{3.80}$$

The first term on the right is due to joule heating and its time averaged value is

$$< p_d >= \frac{1}{2}\hat{\mathbf{E}}.\hat{\mathbf{J}}^* \tag{3.81}$$

The time-averaged stored energy density in the electric and magnetic fields is

$$< w_E >= \frac{1}{4}\epsilon\hat{\mathbf{E}}.\hat{\mathbf{E}}^*, \quad < w_H >= \frac{1}{4}\mu\hat{\mathbf{H}}.\hat{\mathbf{H}}^* \tag{3.82}$$

Hence

$$\nabla.(\hat{\mathbf{E}} \times \hat{\mathbf{H}}^*) = -2 < p_d > -4j\omega[< w_H > - < w_E >] \tag{3.83}$$

and the power flow is

$$\oint_S <\mathbf{S}> .d\mathbf{s} = \frac{1}{2}\oint_S \hat{\mathbf{E}} \times \hat{\mathbf{H}}^*.d\mathbf{s}$$

$$= -\int_V < p_d > dV - 2j\omega \int_V [< w_H > - < w_E >]dV \tag{3.84}$$

This is the complex Poynting equation, which relates the electromagnetic power flowing through a surface S, the joule heat dissipation, and the difference in energy stored in the electric and magnetic fields within the volume V of material. If there is a power source P_s within the material, then this is added to the right-hand side of this equation, which can then be written as

$$P_s = P_d + 2j\omega(W_H - W_E) + P_o \tag{3.85}$$

where

$$P_d = -\int_V < p_d > dV \tag{3.86}$$

$$W_H - W_E = \int_V [< w_H > - < w_E >]dV \tag{3.87}$$

$$P_o = \frac{1}{2}\oint_S \hat{\mathbf{E}} \times \hat{\mathbf{H}}^*.d\mathbf{s} \tag{3.88}$$

Example
A plane wave propagates in the z-direction in free space. The electric field is $\mathbf{E} = 100\ \mathbf{a_x}\ \mathrm{Vm}^{-1}$. Determine the power density and its direction.
  The time average Poynting vector

$$<\mathbf{S}>= \mathcal{R}_e\frac{1}{2}[\hat{\mathbf{E}} \times \hat{\mathbf{H}}^*] \tag{3.89}$$

gives the time average power density and its direction. The characteristic impedance is

$$Z_o = \frac{E_x}{H_y} \tag{3.90}$$

where $Z_o = 377\ \Omega$ is the characteristic impedance of free space. Thus

$$E_x H_y = E_x^2/Z_o \tag{3.91}$$

Hence,

$$< \mathbf{S} >= \frac{1}{2}E_xH_y\mathbf{a}_z = \frac{1}{2}E_x^2/Z_o\mathbf{a}_z \tag{3.92}$$

and

$$< \mathbf{S} >= 10^4/754\ \mathbf{a}_z = 13.26\ \mathbf{a}_z,\quad Wm^{-2} \tag{3.93}$$

Hence, the power flow density is 13.26 Wm^{-2} in the z-direction of propagation.

3.3.4   Relaxation dependence

In general the conductivity, permittivity and permeability may be complex due to relaxation dependence and become, respectively

$$\sigma = \sigma' + j\sigma'',\quad \epsilon = \epsilon' - j\epsilon'',\quad \epsilon^* = \epsilon' + j\epsilon'',\quad \mu = \mu' - j\mu'' \tag{3.94}$$

The complex Poynting equation is then

$$\oint_S < \mathbf{S} > .d\mathbf{s} = \frac{1}{2}\int_V [(-\sigma'\hat{\mathbf{E}}.\hat{\mathbf{E}}^* - \omega\epsilon''\hat{\mathbf{E}}.\hat{\mathbf{E}}^* - \omega\mu''\hat{\mathbf{H}}.\hat{\mathbf{H}}^*)$$
$$+ j(-\sigma''\hat{\mathbf{E}}.\hat{\mathbf{E}}^* + \omega(\epsilon'\hat{\mathbf{E}}.\hat{\mathbf{E}}^* - \mu'\hat{\mathbf{H}}.\hat{\mathbf{H}}^*))]dV \tag{3.95}$$

If the field contains a source of power P_s then Poynting's equation becomes

$$\oint_S < \mathbf{S} > .d\mathbf{s} = \frac{1}{2}\oint_S \hat{\mathbf{E}} \times \hat{\mathbf{H}}^* .d\mathbf{s}$$
$$= P_s - \int_V p_d dV - 2j\omega \int_V [< W_H > - < W_E >]dV \tag{3.96}$$

or

$$P_s = \int_V p_d dV + 2j\omega \int_V [< W_H > - < W_E >]dV + \frac{1}{2}\oint_S \hat{\mathbf{E}} \times \hat{\mathbf{H}}^* .d\mathbf{s} \tag{3.97}$$

Taking into account relaxation effects this becomes

$$P_s = \frac{1}{2}\int_V [\sigma'\hat{\mathbf{E}}.\hat{\mathbf{E}}^* + \omega(\epsilon''\hat{\mathbf{E}}.\hat{\mathbf{E}}^* + \mu''\hat{\mathbf{H}}.\hat{\mathbf{H}}^*)$$
$$+ j(\sigma''\hat{\mathbf{E}}.\hat{\mathbf{E}}^* + \omega(\mu'\hat{\mathbf{H}}.\hat{\mathbf{H}}^* - \epsilon'\hat{\mathbf{E}}.\hat{\mathbf{E}}^*))]dV + \frac{1}{2}\oint_S \hat{\mathbf{E}} \times \hat{\mathbf{H}}^* .d\mathbf{s} \tag{3.98}$$

If the material contains superconducting currents obeying London theory the power becomes

$$P_s = \frac{1}{2}\int_V [\sigma_n'E^2 + \omega(\epsilon''E^2 + \mu''H^2)$$
$$+ j\omega(\sigma_n''E^2\omega^{-1} + \Lambda J_s^2 + \mu'H^2 - \epsilon'E^2)]dV + \frac{1}{2}\oint_S \hat{\mathbf{E}} \times \hat{\mathbf{H}}^* .d\mathbf{s} \tag{3.99}$$

where $\Lambda = \mu\lambda_L^2$. Since μ is complex, $\mu = \mu' - j\mu''$, this leads to

$$P_s = \frac{1}{2}\int_V [\sigma_n' E^2 + \omega(\epsilon'' E^2 + \mu'' H^2 + \mu'' \lambda_L^2 J_s^2)$$

$$+ j\omega(\sigma_n'' E^2 \omega^{-1} + \mu' \lambda_L^2 J_s^2 + \mu' H^2 - \epsilon' E^2)]dV + \frac{1}{2}\oint_S \hat{E} \times \hat{H}^* .ds \quad (3.100)$$

3.4   Impedance

Defining the wave impedance as

$$Z(\omega) = \frac{\hat{E}}{\hat{H}}, \quad \hat{H} = \frac{\hat{E}}{Z(\omega)}, \quad \hat{E} = \hat{H}Z(\omega) \quad (3.101)$$

then since $\hat{H}.\hat{H}^* = H_o^2$ and $\hat{E}.\hat{E}^* = E_o^2$

$$<P_s> = \frac{H_o^2}{2}Re[Z(\omega)] = \frac{E_o^2}{2}Re[1/Z(\omega)] \quad (3.102)$$

3.5   Complex voltage and current

Equations similar to Poynting's vector can be derived by considering complex current and voltage dissipating power in a lossy circuit. In this, we assume the quasistatic approximation with negligible radiation losses in which voltage v and current i can be determined. By definition, the electrical power supplied is vi where

$$v = v_o cos(\omega t + \theta), \quad i = i_o cos(\omega t + \phi) \quad (3.103)$$

$$v = Re[v_o e^{j(\omega t + \theta)}], \quad i = Re[i_o e^{j(\omega t + \phi)}] \quad (3.104)$$

$$v = Re[\hat{v}e^{j\omega t}], \quad i = Re[\hat{i}e^{j\omega t}] \quad (3.105)$$

where

$$\hat{v} = v_o e^{j\theta}, \quad \hat{i} = i_o e^{j\phi} \quad (3.106)$$

The instantaneous power dissipated is

$$P_i = vi = Re[v_o e^{j(\omega t + \theta)}]Re[i_o e^{j(\omega t + \phi)}] \quad (3.107)$$

$$P_i = v_o i_o cos(\omega t + \theta)cos(\omega t + \phi). \quad (3.108)$$

Using

$$cos[(A + B)/2]cos[(A - B)/2] = (cosA + cosB)/2 \quad (3.109)$$

$$\omega t + \theta = (A + B)/2, \quad \omega t + \phi = (A - B)/2 \quad (3.110)$$

gives

$$A = 2\omega t + \theta + \phi, \quad B = \theta - \phi \quad (3.111)$$

$$P_i = v_o i_o(cos A + cos B)/2 \quad (3.112)$$

$$P_i = v_o i_o [\cos(2\omega t + \theta + \phi) + \cos(\theta - \phi)]/2 \tag{3.113}$$

The time average of the instantaneous power flow is

$$< P_i > = \frac{1}{T} \int_0^T P_i dt = \frac{v_o i_o}{2} \cos(\theta - \phi) = Re[\frac{v_o i_o}{2} e^{j(\theta - \phi)}] \tag{3.114}$$

since the time average of the first cosine term is zero. Hence,

$$< P_i > = < vi > = \frac{1}{2} Re[\hat{v}\hat{i}^*] = \frac{1}{2} Re[\hat{v}^*\hat{i}] \tag{3.115}$$

where

$$\hat{v} = v_o e^{j\theta}, \quad \hat{v}^* = v_o e^{-j\theta}, \quad \hat{i} = i_o e^{j\phi}, \quad \hat{i}^* = i_o e^{-j\phi} \tag{3.116}$$

3.5.1   Alternative analysis

Equation (3.105) can be expressed as

$$v = (\hat{v}e^{j\omega t} + \hat{v}^* e^{-j\omega t})/2, \quad i = (\hat{i}e^{j\omega t} + \hat{i}^* e^{-j\omega t})/2 \tag{3.117}$$

Proof:

$$i = (\hat{i}e^{j\omega t} + \hat{i}^* e^{-j\omega t})/2 \tag{3.118}$$

$$= i_o[\cos(\omega t + \phi) + j\sin(\omega t + \phi)]/2 + i_o^*[\cos(\omega t + \phi) - j\sin(\omega t + \phi)]/2 \tag{3.119}$$

$$= i_o \cos(\omega t + \phi), \quad qed \tag{3.120}$$

The instantaneous power dissipated is

$$P_i = vi = (\hat{v}e^{j\omega t} + \hat{v}^* e^{-j\omega t})(\hat{i}e^{j\omega t} + \hat{i}^* e^{-j\omega t})/4 \tag{3.121}$$

$$= (\hat{v}\hat{i}e^{2j\omega t} + \hat{v}\hat{i}^* + \hat{v}^*\hat{i} + \hat{v}^*\hat{i}^* e^{-2j\omega t})/4 \tag{3.122}$$

The time average of the instantaneous power flow is

$$< P_i > = \frac{1}{T} \int_0^T P_i dt = (\hat{v}\hat{i}^* + \hat{v}^*\hat{i})/4 \tag{3.123}$$

since the time average of the first and fourth periodic terms are zero. Hence, assuming $\hat{v}\hat{i}^* = \hat{v}^*\hat{i}$

$$< P_i > = Re(\hat{v}\hat{i}^*)/2 = Re(\hat{v}^*\hat{i})/2 \tag{3.124}$$

3.5.2   Impedance

Define the impedance as

$$Z(\omega) = \frac{\hat{v}}{\hat{i}}, \quad \hat{i} = \frac{\hat{v}}{Z(\omega)}, \quad \hat{v} = \hat{i}Z(\omega) \tag{3.125}$$

Then, since $\hat{i}\hat{i}^* = i_o^2$ and $\hat{v}\hat{v}^* = v_o^2$

$$< P_i >=< vi >= \frac{i_o^2}{2} Re[Z(\omega)] = \frac{v_o^2}{2} Re[1/Z(\omega)] \tag{3.126}$$

3.6 Power dissipation in an LCR circuit

For a general series L, C, R circuit the power applied to the circuit = the power absorbed by the circuit

$$vi = v_L i_L + v_C i_C + v_R i_R \tag{3.127}$$

where $I_L = i_R$. The basic equations for L, C and R are

$$v_L = L\frac{di_L}{dt}, \quad i_C = C\frac{dv_C}{dt}, \quad v_R = i_R R \tag{3.128}$$

The power dissipated in each component

$$i_L v_L = i_L L\frac{di_L}{dt} = \frac{d}{dt}(Li_L^2/2), \quad i_C v_C = Cv_C\frac{dv_C}{dt} = \frac{d}{dt}(Cv_C^2/2), \quad i_R v_R = i_R^2 R \tag{3.129}$$

$$[Note : If \; ydy = dy^2/2, \; then \; \int ydy = \int dy^2/2, \; i.e. \; y^2/2 = y^2/2] \tag{3.130}$$

The power dissipated is then

$$vi = \frac{d}{dt}(Li_L^2/2) + \frac{d}{dt}(Cv_C^2/2) + i_R^2 R \tag{3.131}$$

This immediately shows that the energy stored in the E and H fields is

$$W_H = \frac{1}{2}Li_L^2, \quad W_E = \frac{1}{2}Cv_C^2 \tag{3.132}$$

An example of Poynting vector and dc Power flow in a co-axial cable is given by Parton *et al.* [20].

Chapter 4
The Skin Effect – introduction

4.1 Introduction

The solution of Maxwell's equations for periodic electromagnetic fields incident on conducting materials gives rise to modified wave equations (Helmholtz equations) (2.50) and (2.54). These equations yield periodic fields resulting from the second derivative and diffusion fields due to the first derivative term. The result combines to give periodic electromagnetic amplitudes, which decrease exponentially from the surface as the wave penetrates into the conductor. Because the fields $\mathbf{E}, \mathbf{H}, \mathbf{J}$ and vector potential $\mathbf{A}$ obey similar Helmholtz equations, these fields will all be expected to diffuse from the surface with amplitudes decreasing smoothly as they penetrate the conductor. The depth of penetration is found to decrease as the frequency increases. At very high frequencies in the region of microwaves, the depth of penetration is of the order of micrometers. The variation of the depth of penetration of the EM fields is referred to as *The Skin Effect*. At microwave frequencies, the skin depth may be less than one micrometer. Hence, only a thin film of high conductivity silver or gold coated onto a rigid substrate is necessary for low-loss microwave applications. Since the microwaves only penetrate into the silver, a less costly substrate material can be used. Although, as we show further on (4.2) high conductivity (low resistivity) metals lead to lower skin depth. This effect is compensated for by the low loss in these conductors.

Although the term *Skin Effect* was originally intended to refer to the shallow depth of wave penetration at high frequencies, it tends to be used more generally to include any depth of penetration if the cause is the same as for microwaves. Hence, even at low frequencies, the Skin Effect can be important. In copper at 50 Hz, the penetration depth is $\delta = 9.28$ mm and $\delta = 8.47$ mm at 60 Hz. Thus, in high power applications, it is a waste of copper to increase the radius above about 1 cm and hollow metal tubes can be used. The skin depth is also inversely proportional to the magnetic permeability μ_r. Hence, if conductors are employed that have high μ_r, steel, for example, then the skin depth will be reduced further as shown in Figure 4.3.

4.2 Skin Effect – a brief history

The notion of the Skin Effect essentially follows closely that of the developments in electromagnetism and radio engineering in particular, [45,60,61,69,70]. Although

James Clerk Maxwell had earlier determined the equations for the penetration of an EM field into a cylindrical conductor, the term *Skin Effect* was not used until later. See references [3] and footnotes by J.J. Thomson and [62]. Some of the earliest experiments on the Skin Effect were carried out by Sir Ambrose Fleming [63]. Later, Arthur Kennelly and his co-workers used a large copper wire meander for measuring the Skin Effect in solid round copper wires [65]. Kennelly's work extended earlier measurements performed by Fleming [63]. These experimental measurements confirmed that if the current is alternating the solution of Maxwell's equations shows that inside the conductor the EM fields obey a diffusion law. The current density decreases towards the centre of the conductor and decreases more sharply as the frequency increases. Thus, because the cross-sectional area of the current path is reduced, the *high frequency resistance* of the conductor increases with frequency according to the well-known result, $R_{ac} \propto \sqrt{f}$; a process generally referred to as the *Skin Effect*. This leads to increased losses as the frequency increases. This was very important in the early years of radio engineering and practical details were discussed widely in the literature [45].

During the subsequent years, there has been a large volume of literature on the theory and modelling of the Skin Effect [41,66]. In comparison, there seems to have been very few papers concerned with directly measuring the Skin Effect in simple conductors particularly at low frequencies. But MacDougal set up some nice low-cost experiments for demonstrating the Skin Effect to undergraduates [64]. Authors of theoretical papers frequently refer to the experimental results of Kennelly [67,68]. However, Kennelly's ac resistance measurements were limited to eight data points obtained between 10 kHz and 100 kHz [65].

4.3   General description of the Skin Effect

The mathematical theory does not provide a physical explanation of the diffusion mechanism, only that it is a result of the solution of Maxwell's equations. By comparison, the diffusion of gases is explained by the atoms or molecules moving from a higher concentration to a lower concentration due to random collisions. But what is the mechanism for electric and magnetic field diffusion? A classical physical explanation of the Skin Effect, [45,59,69], is that when alternating current flows in a conductor, the variation in magnetic field induces secondary currents, *Eddy Currents*, which increase with increasing frequency. These oppose the externally applied supply current in accordance with Lenz's law. In the case of cylindrical conductors or wires, the amount of magnetic flux linking the conductor is higher at the centre of the conductor than the surface, Figure 4.1(a). This reduces the supply current in the centre of the conductor, increasing its resistance, which increases further as the frequency increases. In the case of a rectangular conducting bar or strip, the greatest concentration of flux lines is in the centre, decreasing at the corners and the surface. This gives the largest current densities at the surfaces, followed by the corners and the least along the central axis of the conductor as for the cylindrical conductor, Figure 4.1(b). However, in the rectangular case a more detailed analysis shows that the field lines

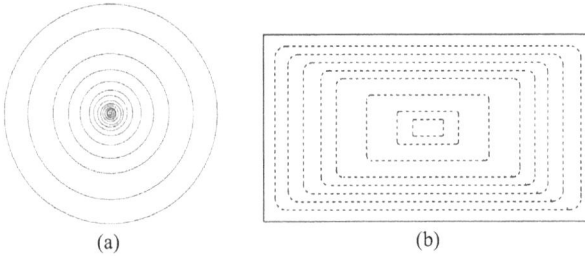

(a)	(b)

Figure 4.1   Flux lines in (a) cylindrical conductor, (b) rectangular conductor. The field line patterns in the corners are more complex than shown.

follow complex patterns at the corners than is shown in this figure. This is not the case for the cylindrical geometry which is more ideal for analytical purposes.

These interpretations may also be considered in terms of the wire self-inductance L. The wire self-inductance is defined as the ratio of the magnetic flux Ψ linking the conductor to the current I, that is, $L = \Psi/I$. If the flux linking the conductor is less at the surface compared with inside the conductor, then the self-inductance is higher inside the conductor compared with outside. The reactance, ωL, is therefore higher inside the conductor and the current is redistributed towards the surface. Again, the effect is more significant as the frequency increases.

If the mean free path of the conduction electrons l_n approaches the field penetration depth then the classical Skin Effect theory breaks down and the nature of the penetration is referred to as the *Anomalous Skin Effect* [72]. This normally occurs at high frequencies since, as mentioned, the skin depth decreases with increasing frequency and eventually $\delta \to l$. In addition, the atomic and electronic structure of solid surfaces may deviate considerably from the bulk properties. In the classical treatment, Maxwell's equations do not consider the microscopic nature of matter, only macroscopic fields averaged over a large number of atomic dimensions.

In general, it is not possible to obtain simple analytical equations for the Skin Effect in an arbitrary shaped conductor such as for example complex shaped transformer laminations or complex geometries in microwave circuits. For such problems the diffusion equations are solved using field plots and numerical methods [20,71,110]. Analytical solutions can, however, be obtained for simple geometries such as cylindrical wire conductors and rectangular conductors. First, however, we consider a simplified method of analysis. This is followed by a more detailed analysis using Bessel functions.

4.4   Conducting half-space

To illustrate the Skin Effect, consider a good conductor such as copper where $\frac{\sigma}{\omega\epsilon} \geq 10$. The transverse electromagnetic (TEM) wave travelling in the z-direction is incident on the conductor, which extends to infinity in the positive x, y and z dimensions, that is, half-space. The electric field is defined by $E_x = E_o \cos(\omega t)$. Inside the conductor this becomes (c.f. (2.84))

$$E_x = E_o e^{-(\alpha + j\beta)z} \cos(\omega t) \tag{4.1}$$

For a good conductor we found that $\alpha = \beta = 1/\delta$ where

$$\delta = 1/\alpha = \sqrt{\frac{2}{\omega\mu\sigma}} \tag{4.2}$$

Hence, we can write

$$E_x = E_o e^{-z/\delta} \cos(\omega t - z/\delta) \tag{4.3}$$

The penetration depth δ has dimensions of metres and is known as the *Skin Depth*. This corresponds to the distance the wave propagates before its amplitude decays to e^{-1} (about 37 per cent) of its maximum value, Figure 4.2.

The magnetic field is

$$H_y = \frac{E_o}{|Z|} e^{-z/\delta} \cos(\omega t - z/\delta - \pi/4) \tag{4.4}$$

where for a good conductor the impedance is

$$Z = \frac{1+j}{\sqrt{2}} \sqrt{\omega\mu/\sigma} = \frac{1+j}{\sigma\delta} \tag{4.5}$$

which gives $\phi = 45°$. The current density is taken as $J_x = \sigma E_x$. Hence this becomes,

$$J_x = \sigma E_o e^{-z/\delta} \cos(\omega t - z/\delta) = J_o e^{-z/\delta} \cos(\omega t - z/\delta) \tag{4.6}$$

Current density penetration into a good conductor

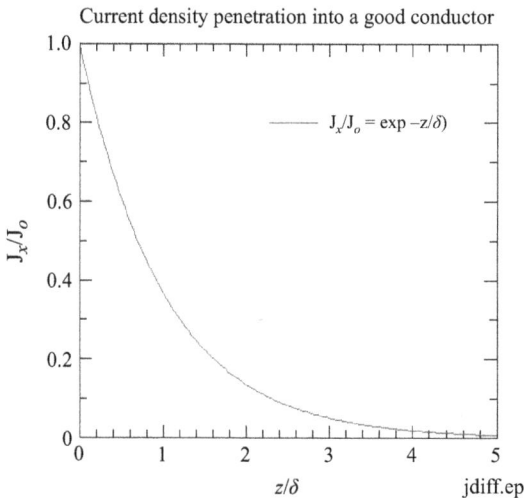

Figure 4.2   Current density variation with distance into a good conductor

Theoretical skin depths at 300K

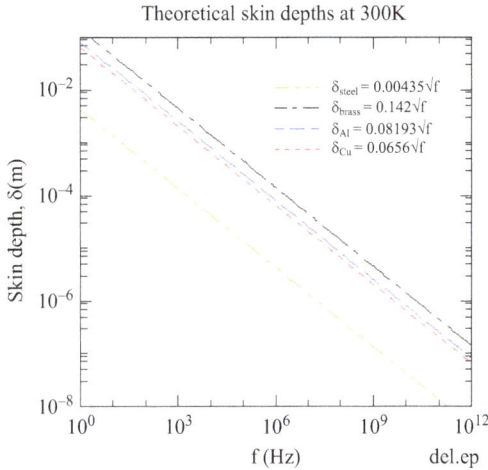

Figure 4.3 Skin depths for a number of metals

where J_o is the current density at the surface of the conductor where $z = 0$. The current density therefore diffuses into the conductor decreasing exponentially with distance from J_o at the surface ($z = 0$) to $J_o e^{-1} = 0.368 J_o$ at $z = \delta$. From (4.2), the skin depth decreases with increasing frequency, so that the Skin Effect becomes particularly important at very high frequencies, Figure 4.3. In this case, the decreasing skin depth for a given conductivity, leads to an enhanced surface impedance as shown by (4.5).

4.5 Approximate methods

By assuming that most of the ac resistance increase occurs for one skin depth then the Skin Effect can be used to estimate the increase in resistance due to EM diffusion. This analysis is approximate because the resistance decreases exponentially from the surface of the conductor. One skin depth accounts for 67 per cent of this resistance change from the surface, $z = 0$ to $z = \delta$. A more detailed analysis requires the solution of the Helmholtz equations for the particular conductor geometry and boundary conditions concerned. This is considered further on in the text.

4.5.1 Cylindrical wire

Consider first a cylindrical conducting wire radius a, length l and skin depth δ, Figure 4.4(a). The d.c. resistance is

$$R_{dc} = \frac{\rho l}{A} \tag{4.7}$$

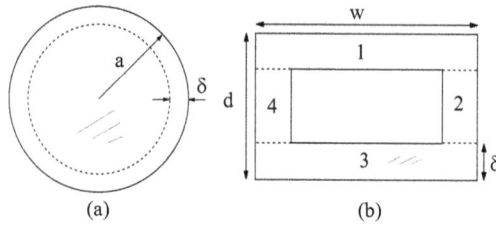

Figure 4.4   Sections through (a) circular conductor and (b) rectangular conductor.
Both conductors have length l perpendicular to the page

where A is the cross-sectional area of the conducting region. For the full conductor $A_1 = \pi a^2$. For the full conductor, less skin depth $A_2 = \pi(a - \delta)^2$. The area for a one skin depth conducting path is a tube with cross-sectional area

$$A_d = A_1 - A_2 = \pi[a^2 - (a - \delta)^2] \tag{4.8}$$

The ac resistance becomes

$$R_{ac} = \frac{\rho l}{A_d} = \frac{\rho l/(\pi a^2)}{1 - (\frac{a-\delta}{a})^2} \tag{4.9}$$

or

$$\frac{R_{ac}}{R_{dc}} = \frac{1}{1 - (\frac{a-\delta}{a})^2} = \frac{1}{1 - (1 - \delta/a)^2} \tag{4.10}$$

Hence, if $\delta = a$ then $R_{ac} = R_{dc}$. If $\delta \ll a$ then $R_{ac} \gg R_{dc}$. See Figure 4.5

4.5.2   Rectangular conductor

Rectangular conductor geometries occur in thin film circuits, microwave striplines and low-frequency power transmission lines. Consider a rectangular conductor with width w, depth d and length l, Figure 4.4(b). Assume that the conducting region is due to only a skin depth δ. In this figure, the conducting sectional areas 1–4 clockwise from the top are

$$A_r = w\delta + (d - 2\delta)\delta + w\delta + (d - 2\delta)\delta = 2\delta(w + d - 2\delta) \tag{4.11}$$

The ac resistance due to one skin depth becomes

$$R_{ac} = \frac{\rho l}{2\delta(w + d - 2\delta)} \tag{4.12}$$

or

$$\frac{R_{ac}}{R_{dc}} = \frac{wd}{2\delta(w + d - 2\delta)} \tag{4.13}$$

If $\delta = w/2 = d/2$ then $R_{ac} = R_{dc}$. If $\delta \ll w \ll d$ then $\frac{R_{ac}}{R_{dc}} = \frac{wd}{2\delta(w+d)}$.

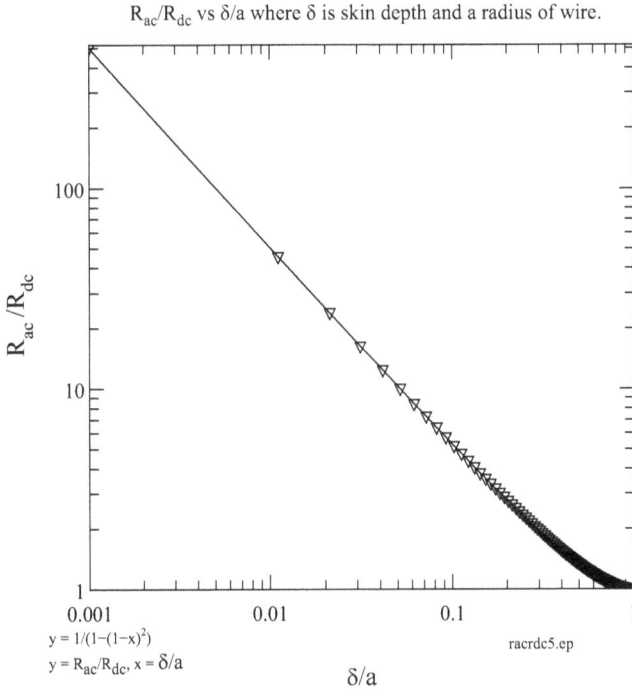

R_{ac}/R_{dc} vs δ/a where δ is skin depth and a radius of wire.

$y = 1/(1-(1-x)^2)$
$y = R_{ac}/R_{dc}, \; x = \delta/a$

racrdc5.ep

Figure 4.5   Variation of ac resistance R_{ac}/R_{dc} with skin depth δ/a

4.5.3   Tubular conductor

Consider a cylindrical tubular conductor outside radius a, internal radius b, length and skin depth δ.

The area for a one skin depth conducting path is a tube with cross-sectional area

$$A_d = \pi[a^2 - (a - \delta)^2] = \pi(2a\delta - \delta^2) \tag{4.14}$$

The ac resistance becomes

$$R_{ac} = \frac{\rho l}{A_d} = \frac{\rho l}{\pi(2a\delta - \delta^2)} \tag{4.15}$$

The dc resistance is

$$R_{dc} = \frac{\rho l}{\pi(a^2 - b^2)} \tag{4.16}$$

Hence, the ratio of ac to dc resistance of the tube for one skin depth is

$$\frac{R_{ac}}{R_{dc}} = \frac{a^2 - b^2}{2a\delta - \delta^2} \tag{4.17}$$

4.6   Methods of reducing Skin Effect

One method of reducing the Skin Effect in a conductor is to use several parallel wires insulated from each other, twisted together and shorted at each end. The total resistance is then approximately $R_T = R/n$, where R is the resistance of one wire and n the number of wires. This arrangement is known as *litz* wire or 'Litzendraht' conductor, and is useful for frequencies up to about 500 kHz. The advantages become less at frequencies beyond 2 MHz due to irregularities in the twisted wires and capacitance between the strands, [45] p. 37. Generally stranded conductors have a large surface area than single conductors, which reduces the Skin Effect. In printed circuit boards (PCBs) thicker copper layers or multiple layers are used to decrease the conducting plane resistance and Skin Effect. A further method recently reported suggests reducing the Skin Effect with a micro metre-scale gridded fibre structure with currents arranged in a checkered pattern [50].

Although the skin depth is proportional to the resistivity of the conductor (see (4.2)) increasing the resistivity will increase the power losses. Hence low resistivity high conductivity metals such as copper or silver are used to keep losses low. Equation (4.2) also shows that the skin depth is inversely proportional to the wire permeability and frequency. Hence, low permeability materials are chosen. Higher frequencies clearly lead to a decrease in skin depth. However, the choice of frequency depends on many other factors, so Skin Effect may not be the only consideration.

Chapter 5
Cylindrical conductor – axial alternating current

5.1 Introduction

An ac generator supplies a constant current $I = I_o e^{j\omega t}$ parallel to the z-axis of a solid cylindrical conductor or wire, length l, radius a and uniform resistivity ρ. In the case of a steady state (dc), the current is distributed uniformly across the section of the conductor, and the current density is simply given by $J = I/(\pi a^2)$ [75]. For alternating currents, the current density is no longer uniform but is a function of the conductor radius r and can be described by $J(r,t) = J(r)e^{j\omega t}$. The following analysis is primarily concerned with the current flow within the conductor. Current leakage and flow outside the conductor are not explicitly included. However, some consideration is given to the external inductance, which can have a large effect on the impedance.

5.2 The electric and magnetic fields from Maxwell's equations

The electric and magnetic fields are determined from Maxwell's equations. The analysis initially leads to the *Helmholtz Wave Equations or Equations of Telegraphy* which are the sum of wave and diffusion equations. These finally give, for the E and H fields, zero-order and first-order Bessel functions of the first kind, respectively, with complex arguments. The Helmholtz Wave Equations were derived previously, page 32, (2.50) and (2.54) repeated here as

$$\nabla^2 \mathbf{E} = \mu\sigma \frac{\partial \mathbf{E}}{\partial t} + \mu\epsilon \frac{\partial^2 \mathbf{E}}{\partial t^2} + \nabla(\rho_v/\epsilon) \tag{5.1}$$

$$\nabla^2 \mathbf{H} = \mu\sigma \frac{\partial \mathbf{H}}{\partial t} + \mu\epsilon \frac{\partial^2 \mathbf{H}}{\partial t^2} \tag{5.2}$$

For conductors like copper, the ratio of conductivity to permittivity is ($\sigma/\epsilon_o \approx 10^{19}$). Therefore, the second derivative term for H_ϕ in (5.2) is neglected, and for good conductors, the Helmholtz equation for the magnetic field becomes only a diffusion equation

$$\nabla^2 \mathbf{H} = \mu\sigma \frac{\partial \mathbf{H}}{\partial t} \tag{5.3}$$

Generally, for cylindrical conductors the components of the vector Laplacian are [6] (p. 92)

$$\nabla^2 \mathbf{S}_\rho = \nabla^2 S_\rho - \frac{S_r}{\rho^2} - \frac{2}{\rho^2} \frac{\partial S_\phi}{\partial \phi} \tag{5.4}$$

$$\nabla^2 \mathbf{S}_\phi = \nabla^2 S_\phi - \frac{S_\phi}{\rho^2} + \frac{2}{\rho^2} \frac{\partial S_\rho}{\partial \phi} \tag{5.5}$$

$$\nabla^2 \mathbf{S}_z = \nabla^2 S_z \tag{5.6}$$

5.2.1 Electric field

The electric field is in the z-direction and the scalar equation (with $r = \rho$) is [20] (p. 256)

$$\nabla^2 E_z = \frac{1}{r} \frac{\partial}{\partial r} \left(r \frac{\partial E_z}{\partial r} \right) + \frac{1}{r^2} \frac{\partial^2 E_z}{\partial \phi^2} + \frac{\partial^2 E_z}{\partial z^2} \tag{5.7}$$

With no variation of E_z with ϕ or z then

$$\nabla^2 E_z = \frac{1}{r} \frac{\partial}{\partial r} \left(r \frac{\partial E_z}{\partial r} \right) = \frac{1}{r} \frac{\partial E_z}{\partial r} + \frac{\partial^2 E_z}{\partial r^2} = \mu\sigma \frac{\partial \mathbf{E}}{\partial t} \tag{5.8}$$

Hence,

$$\frac{\partial^2 E_z}{\partial r^2} + \frac{1}{r} \frac{\partial E_z}{\partial r} - \mu\sigma \frac{\partial E_z}{\partial t} = 0 \tag{5.9}$$

5.2.2 Magnetic field

In this case for H_ϕ with $r = \rho$ no variation with ϕ and from (5.5) the vector Laplacian is

$$\nabla^2 \mathbf{H}_\phi = \nabla^2 H_\phi - \frac{H_\phi}{r^2} \tag{5.10}$$

The scalar Laplacian for H_ϕ is [20]

$$\nabla^2 H_\phi = \frac{1}{r} \frac{\partial}{\partial r} \left(r \frac{\partial H_\phi}{\partial r} \right) + \frac{1}{r^2} \frac{\partial^2 H_\phi}{\partial \phi^2} + \frac{\partial^2 H_\phi}{\partial z^2} \tag{5.11}$$

With no variation of H_ϕ with ϕ or z then

$$\nabla^2 H_\phi = \frac{1}{r} \frac{\partial}{\partial r} \left(r \frac{\partial H_\phi}{\partial r} \right) = \frac{1}{r} \frac{\partial H_\phi}{\partial r} + \frac{\partial^2 H_\phi}{\partial r^2} \tag{5.12}$$

Hence,

$$\nabla^2 H_\phi = \frac{\partial^2 H_\phi}{\partial r^2} + \frac{1}{r} \frac{\partial H_\phi}{\partial r} - \frac{H_\phi}{r^2} = \mu\sigma \frac{\partial H_\phi}{\partial t} \tag{5.13}$$

An alternative solution is given in Appendix.

5.3   Sine waves

For sine waves, $\partial/\partial t = j\omega$ (5.13) and (5.9) become, respectively

$$\frac{\partial^2 H_\phi}{\partial r^2} + \frac{1}{r}\frac{\partial H_\phi}{\partial r} - \left(\frac{1}{r^2} + j\omega\mu\sigma\right)H_\phi = 0 \tag{5.14}$$

$$\frac{\partial^2 E_z}{\partial r^2} + \frac{1}{r}\frac{\partial E_z}{\partial r} - j\omega\mu\sigma E_z = 0 \tag{5.15}$$

Putting

$$u = mrj^{3/2}, \quad m = \sqrt{(\omega\mu\sigma)} \tag{5.16}$$

$$\frac{\partial^2 H_\phi}{\partial u^2} + \frac{1}{u}\frac{\partial H_\phi}{\partial u} + (1 - 1/u^2)H_\phi = 0 \tag{5.17}$$

$$\frac{\partial^2 E_z}{\partial u^2} + \frac{1}{u}\frac{\partial E_z}{\partial u} + E_z = 0 \tag{5.18}$$

Bessel's equation order v is

$$\frac{d^2 y}{dx^2} + \frac{1}{x}\frac{dy}{dx} + (1 - v^2/x^2)y = 0 \tag{5.19}$$

Comparing this with (5.17) and (5.18) gives $v = 1$ and $v = 0$, respectively. Hence the H and E fields are given by first-order and zero-order Bessel equations, respectively, with complex argument u.

From (5.106)

$$H_\phi = \frac{I_o}{2\pi a}\frac{J_1(u)}{J_1(u_a)} = kJ_1(u) \tag{5.20}$$

where J_1 is a first-order Bessel function with complex argument u given by (5.16) and u_a with $r = a$. I_o is the current amplitude and $k = I_o/(2\pi a J_1(u_a))$. Substituting (5.20) in (5.17)

$$y = J_1'' + J_1'/u + J_1 - J1/u^2 \tag{5.21}$$

Now

$$J_1' = J_o - J_1/u, \quad J_1'' = J_o' - J_1''/u + J_1/u^2 \tag{5.22}$$

Thus,

$$y = J_o' - J_1'/u + J_1/u^2 + J_1'/u + J_1 - J_1/u^2 = J_o' - J_1 \tag{5.23}$$

But $J_o' = -J_1$. Hence, $y = 0$ as expected.

5.4   The current density

For a good conductor with current flowing in the z-direction, and assuming the displacement current is zero, then

$$J_z = \sigma E_z, \quad E_z = \rho J_z \tag{5.24}$$

Equation (5.9) becomes

$$\frac{\partial^2 J_z}{\partial r^2} + \frac{1}{r}\frac{\partial J_z}{\partial r} - \mu\sigma\frac{\partial J_z}{\partial t} = 0 \tag{5.25}$$

5.4.1   Sine waves

For sine waves, $J_z(r,t) = J_z e^{j\omega t}$ and $\partial/\partial t = j\omega$, (5.25) becomes

$$\frac{\partial^2 J_z}{\partial r^2} + \frac{1}{r}\frac{\partial J_z}{\partial r} - j\omega\mu\sigma J_z = 0 \tag{5.26}$$

Putting

$$u = mrj^{3/2}, \quad m = \sqrt{(\omega\mu\sigma)} \tag{5.27}$$

$$\frac{\partial^2 J_z}{\partial u^2} + \frac{1}{u}\frac{\partial J_z}{\partial u} + J_z = 0 \tag{5.28}$$

Bessel's modified equation order v for a function $Z_v(u)$ is defined by

$$\frac{\partial^2 Z_v(u)}{\partial u^2} + \frac{1}{u}\frac{\partial Z_v(u)}{\partial u} + (u^2 - v^2)Z_v(u) = 0 \tag{5.29}$$

Comparing this with (5.28) gives $v = 0$. Hence J and E fields are given by a zero-order Bessel equation with complex argument u, where $u = mrj\sqrt{j}$. The general solution is given by

$$Z_v(u) = AJ_v(u) + BN_v(u) \tag{5.30}$$

where $J_v(u)$ and $N_v(u)$ are Bessel functions of the first and second kind, respectively, with order v. $J_v(u)$ are obtained from a series solution of the Bessel equation [6] where

$$J_v(u) = \sum_{s=0}^{\infty} \frac{(-1)^s}{s!(v+s)!}(u/2)^{v+2s} \tag{5.31}$$

$$J_v(u) = (u/2)^v \left[\frac{1}{v!} - \frac{(u/2)^2}{1!(v+1)!} + \frac{(u/2)^4}{2!(v+2)!} - \frac{(u/2)^6}{3!(v+3)!} + \cdots \right. \tag{5.32}$$

$$= \frac{1}{\pi}\int_0^\pi \cos(n\theta - u\sin\theta)\, d\theta \tag{5.33}$$

$$N_v(u) = \frac{J_v(u)\cos v\pi - J_{-v}(u)}{\sin n\pi} \tag{5.34}$$

The Bessel functions of order zero is

$$J_o(u) = 1 - \frac{(u/2)^2}{(1!)^2} + \frac{(u/2)^4}{(2!)^2} - \frac{(u/2)^6}{(3!)^2} + \cdots$$

$$= \frac{1}{\pi}\int_0^\pi \cos(u\sin\theta)\, d\theta \Rightarrow 1, \quad u \Rightarrow 0 \tag{5.35}$$

The Bessel functions of order one is

$$J_1(u) = \frac{u}{2} - \frac{(u/2)^3}{1!2!} + \frac{(u/2)^5}{2!3!} - \frac{(u/2)^7}{3!4!} + \cdots \Rightarrow 0, \quad u \Rightarrow 0 \tag{5.36}$$

Expansion of the series shows that at the origin $J_v(u)$ is zero but $N_v(u)$ is infinite. In our case (5.28), the order $v = 0$. A physical solution for this case is therefore

$$J(u) = AJ_o(u), \quad J(a) = AJ_o(u_a) \tag{5.37}$$

and the constant A has to be determined. At the boundary of the conductor, $r = a$, where $u_a = amj\sqrt{j}$. Taking the ratio of these two equations eliminates A. Hence,

$$J(u, t) = J(a)\frac{J_o(u)}{J_o(u_a)}e^{j\omega t} \tag{5.38}$$

Equation (5.28), then becomes

$$\frac{\partial^2 J_o}{\partial u^2} + \frac{1}{u}\frac{\partial J_o}{\partial u} + J_o = 0 \tag{5.39}$$

Putting this as

$$J_o'' + J_o'/u + J_o = 0 \tag{5.40}$$

where the primes / and // refer to the first and second derivative of u respectively and since $J_o' = -J_1$ and $J_o'' = -(J_o - J_1/u)$, then $y = 0$ as expected.

J_a in (5.38) is not easily measured in practice. We, therefore, consider the following. The ac generator supplies a current $I = I_o e^{j\omega t}$. The current amplitude is then

$$I_o = \int_S \mathbf{J}.d\mathbf{s} = A\int_0^a J_o(u)2\pi r dr \tag{5.41}$$

$$I_o = \frac{2\pi A}{(mj^{3/2})^2}\int_0^{u_a} uJ_o(u)du \tag{5.42}$$

Using the identity $\int uJ_o(u)du = uJ_1(u)$

$$I_o = \frac{j2\pi A}{m^2}u_aJ_1(u_a) \tag{5.43}$$

$$A = \frac{m^2 I_o}{j2\pi u_aJ_1(u_a)} = \frac{mI_o}{j^{3/2}2\pi aJ_1(u_a)} \tag{5.44}$$

Substituting for A in (5.37)

$$J = \frac{j^{3/2}mI_o}{2\pi a}\frac{J_o(u)}{J_1(u_a)} = \frac{I_o}{\pi a^2}\frac{u_a}{2}\frac{J_o(u)}{J_1(u_a)} \tag{5.45}$$

Hence, the current density is

$$J = J_{dc}\frac{u_a}{2}\frac{J_o(u)}{J_1(u_a)}, \quad J_{dc} = \frac{I_o}{\pi a^2} \tag{5.46}$$

For $r = a$

$$J_a = J_{dc}\frac{u_a}{2}\frac{J_o(u_a)}{J_1(u_a)} \tag{5.47}$$

As $\omega \to 0$, $J_o(u) \to 1$, $J_1(u_a) \to u_a/2$ and $J(r) \to \frac{I}{\pi a^2} = \bar{J}(r)$; the average or dc current density, (see following section).

5.4.2   Average current density

The average current density is defined by

$$\bar{J}(r) = \frac{2}{a^2} \int_0^a rJ(r)dr \tag{5.48}$$

Thus, if the current density in the specimen J_{int} is uniform and constant then the average internal current density is $\bar{J} = (2/a^2)J_{int}[a^2/2] = J_{int}$ as expected. Substituting for the current density (5.45) gives

$$\bar{J}(r) = \frac{2}{a^2} \frac{j^{3/2}mI_o}{2\pi aJ_1(u_a)}(mj^{3/2})^{-2} \int_0^{u_a} uJ_o(u)du \tag{5.49}$$

Using the identity

$$\int uJ_o(u)du = uJ_1(u) \tag{5.50}$$

gives

$$\bar{J}(r) = \frac{2}{a^2} \frac{j^{3/2}mI_o}{2\pi aJ_1(u_a)}(mj^{3/2})^{-2} u_a J_1(u_a) \tag{5.51}$$

$$\bar{J}(r) = \frac{j^{3/2}mI_o}{2\pi aJ_1(u_a)} \frac{2}{u_a}J_1(u_a) \tag{5.52}$$

Thus

$$\bar{J}(r) = \frac{I_o}{\pi a^2} = J_{dc} \tag{5.53}$$

The average current density is therefore the total applied current divided by the cross-sectional area of the wire as expected. This is the same as the steady-state dc current density J_{dc}.

5.4.3   Kelvin equations

The current density may also be expressed in terms of Kelvin functions

$$J_o(u) = ber_o(mr) + jbei_o(mr) \tag{5.54}$$

$$J_1(u) = -\frac{J_o(u)}{du} = -\frac{dJ_o(u)}{j^{3/2}d(mr)} = j^2 j^{-3/2}\frac{dJ_o(u)}{d(mr)} = j^{1/2}\frac{dJ_o(u)}{d(mr)} \tag{5.55}$$

Thus,

$$J_1(u_a) = j^{1/2}[ber'_o(ma) + jbei'_o(ma)] \tag{5.56}$$

Hence,

$$J = \left(\frac{j^{3/2}mI_o}{2\pi a}\right) \frac{ber_o(mr) + jbei_o(mr)}{j^{1/2}[ber'_o(ma) + jbei'_o(ma)]} \tag{5.57}$$

$$J = \left(\frac{mI_o}{2\pi a}\right) j \frac{ber_o(mr) + jbei_o(mr)}{[ber'_o(ma) + jbei'_o(ma)]} \tag{5.58}$$

$$J = \left(\frac{mI_o}{2\pi a}\right) \frac{-bei_o(mr) + jber_o(mr)}{[ber'_o(ma) + jbei'_o(ma)]} \tag{5.59}$$

$$J = \left(\frac{mI_o}{2\pi a}\right) \frac{ber_o(mr) + jbei_o(mr)}{[bei'_o(ma) - jber'_o(ma)]} \tag{5.60}$$

where from (5.43)

$$I_o = \frac{-j2\pi Aa}{m}[ber'_o(ma) + jbei'_o(ma)] \tag{5.61}$$

or

$$I_o = \frac{2\pi Aa}{m}[bei'_o(ma) - jber'_o(ma)] \tag{5.62}$$

Rationalising (5.60) gives the real and imaginary components of current density

$$Re(J) = \left(\frac{mI_o}{2\pi a}\right) \frac{ber_o(mr)bei'_o(ma) - bei_o(mr)ber'_o(ma)}{ber'^2_o(ma) + bei'^2_o(ma)} \tag{5.63}$$

$$Im(J) = \left(\frac{mI_o}{2\pi a}\right) \frac{ber_o(mr)ber'_o(ma) + bei_o(mr)bei'_o(ma)}{ber'^2_o(ma) + bei'^2_o(ma)} \tag{5.64}$$

The final current density may be written in terms of the amplitude and phase angle since for any two complex numbers z_1 and z_2 we can divide their moduli and subtract their arguments, that is, $z_1/z_2 = (r_1/r_2) \, arg(\phi_1 - \phi_2)$. Hence,

$$J(t) = \left(\frac{mI_o}{2\pi a}\right) \sqrt{\frac{ber^2_o(mr) + bei^2_o(mr)}{ber'^2_o(ma) + bei'^2_o(ma)}} e^{j(\omega t + \phi)} \tag{5.65}$$

where the phase angle is

$$\phi = \phi_1 - \phi_2 = tan^{-1}\frac{bei_o(mr)}{ber_o(mr)} - tan^{-1}\frac{ber'_o(ma)}{bei'_o(ma)} \tag{5.66}$$

5.5   Axial impedance

Let the axial impedance of the wire be defined by

$$Z(r) = \frac{V(r)}{I} \tag{5.67}$$

where $V(r)$ is the potential difference along the wire and I the total supply current. $V(r)$ is defined by

$$V(r) = -\int_l \mathbf{E}(r).d\mathbf{l} \tag{5.68}$$

where $\mathbf{E}(r)$ is the axial electric field given by

$$-\mathbf{E}(r) = \nabla V_i(r) + \frac{\partial \mathbf{A}(r)}{\partial t} \tag{5.69}$$

$V_i(r)$ is the pd due to the internal properties of the conductor defined by $r \leq a$. $\mathbf{A}(r)$ is the vector magnetic potential arising from the alternating current. This is given by $\mathbf{A}(r) = \mathbf{A}_{int}(r) + \mathbf{A}_{ext}(r)$ with internal and external values, respectively. Substituting (5.69) into (5.68) gives the impedance

$$Z(r) = -\frac{1}{I}\int_l \mathbf{E}(r).d\mathbf{l} = \frac{1}{I}\int_l \left[\nabla V_i(r) + \frac{\partial \mathbf{A}(r)}{\partial t}\right].d\mathbf{l} \tag{5.70}$$

$$Z(r) = \frac{1}{I}\int_l \nabla V_i(r).d\mathbf{l} + \frac{1}{I}\int_l \frac{\partial \mathbf{A}(r)}{\partial t}.d\mathbf{l} \tag{5.71}$$

The first term on the rhs of (5.71) is the impedance due to the internal properties of the conductor. The second term is the contribution from the wire inductance. We consider these contributions in turn.

5.5.1   Internal impedance term

Noting that in the first term in (5.71)

$$\nabla V_i(r) = \frac{\partial V_i(r)}{\partial \mathbf{l}} \tag{5.72}$$

then the first term simply integrates to

$$Z_i(r) = \frac{V_i(r)}{I}, \quad r \leq a \tag{5.73}$$

The current density at a is

$$J = \sigma E_i(r) = \sigma \frac{V_i(r)}{l}, \quad V_i(r) = l\rho J(r) \tag{5.74}$$

Thus,

$$Z_i(r) = \rho l \frac{J(r)}{I} \tag{5.75}$$

From (5.45) with J determined at a

$$J(r)/I = \frac{j^{3/2}m}{2\pi a}\frac{J_o(u_a)}{J_1(u_a)} \tag{5.76}$$

Substitute in (9.21), and since $m = \sqrt{2}/\delta$, then

$$Z_i(r) = \frac{j^{3/2}\sqrt{2}\rho l}{2\pi a\delta}\frac{J_o(u_a)}{J_1(u_a)} \tag{5.77}$$

Since the dc resistance is $R_{dc} = \rho l/(\pi a^2)$ then the internal impedance may be written as

$$Z_i = R_{dc}\frac{j^{3/2}a}{\sqrt{2}\delta}\frac{J_o(u_a)}{J_1(u_a)} \tag{5.78}$$

or

$$Z_i = R_{dc}\frac{u_a}{2}\frac{J_o(u_a)}{J_1(u_a)} \tag{5.79}$$

If $\delta \ll a$ then $2\pi a \delta$ is the cross-sectional area of a tube with thickness δ. The resistance of such a tube is

$$R_1 = \frac{\rho l}{2\pi a \delta} \tag{5.80}$$

As $\omega \to 0$, $Z \to R_{dc}$ as expected. Note that the factor I cancels in the impedance calculation, so that no a priori knowledge of the applied current is required.

5.5.2   Average axial impedance

The axial impedance averaged spatially over the conductor radius is defined by

$$\bar{Z}_i = \frac{2}{a^2} \int_0^a r Z_i(r) dr \tag{5.81}$$

Substituting for (5.79) gives the average axial impedance as equal to the dc resistance

$$\bar{Z}_i = R_{dc} \tag{5.82}$$

5.5.3   Axial impedance and Kelvin functions

The Kelvin functions can be introduced by substituting the current density given by (5.60) into (5.75) to give

$$Z_i = \frac{ml\rho}{2\pi a} \frac{[ber_o(mr) + jbei_o(mr)]}{[bei'_o(ma) - jber'_o(ma)]} \tag{5.83}$$

Substituting $\frac{ml\rho}{2\pi a} = \frac{R_{dc}a}{\sqrt{2}\delta}$ (as in (5.79)) then

$$Z_i = \frac{R_{dc}a}{\sqrt{2}\delta} \frac{[ber_o(mr) + jbei_o(mr)]}{[bei'_o(ma) - jber'_o(ma)]} \tag{5.84}$$

This is similar to Marion and Heald, (5.85), [78] except that the signs are reversed in numerator and denominator. This is because they use an "unconventional" negative sign.

The final impedance may be written in terms of the amplitude and phase angle since as before, for any two complex numbers z_1 and z_2, we can divide their moduli and subtract their arguments, that is, $z_1/z_2 = (r_1/r_2) \arg(\phi_1 - \phi_2)$. Hence,

$$Z_i(r) = |Z_i| e^{j\phi} \tag{5.85}$$

where the amplitude is

$$|Z_i| = \frac{R_{dc}}{\sqrt{2}} \frac{a}{\delta} \sqrt{\frac{ber_o^2(mr) + bei_o^2(mr)}{ber_o'^2(ma) + bei_o'^2(ma)}} \tag{5.86}$$

and the phase angle is

$$\phi = \phi_1 - \phi_2 = tan^{-1}\frac{bei_o(mr)}{ber_o(mr)} - tan^{-1}\frac{ber'_o(ma)}{bei'_o(ma)} \tag{5.87}$$

Also the quadrature components are

$$Re(Z_i) = |Z_i|\cos\phi, \quad Im(Z_i) = |Z_i|\sin\phi \tag{5.88}$$

Kelvin functions are useful for expressing the real and imaginary components yielding the ac resistance, reactance and inductance and allowing these to be compared with experimental measurements. Hence, rationalising (5.84), the real and imaginary parts are

$$Re(Z_i) = \frac{aR_{dc}}{\sqrt{2}\delta} \frac{ber_o(mr)bei'_o(ma) - bei_o(mr)ber'_o(ma)}{ber'^2_o(ma) + bei'^2_o(ma)} \tag{5.89}$$

$$Im(Z_i) = \frac{aR_{dc}}{\sqrt{2}\delta} \frac{ber_o(mr)ber'_o(ma) + bei_o(mr)bei'_o(ma)}{ber'^2_o(ma) + bei'^2_o(ma)} \tag{5.90}$$

The internal inductance is $L_i = Im(Z_i)/\omega$ or

$$L_i = \frac{aR_{dc}}{\omega\sqrt{2}\delta} \frac{ber_o(mr)ber'_o(ma) + bei_o(mr)bei'_o(ma)}{ber'^2_o(ma) + bei'^2_o(ma)} \tag{5.91}$$

In terms of Bessel functions

$$L_i = \frac{Im(Z_i)}{\omega} = \frac{R_{dc}}{\omega} Im\left(\frac{u_a}{2}\frac{J_o(u_a)}{J_1(u_a)}\right) \tag{5.92}$$

In a following chapter, the internal inductance is also calculated by considering the energy in the magnetic field. The following chapter also includes experimental measurements of the internal inductance.

5.6 Appendix

5.6.1 Induced magnetic field

The magnetic field arising from the applied current is obtained from Maxwell's equation

$$curl\ \mathbf{E} = -\frac{\partial\mathbf{B}}{\partial t} \tag{5.93}$$

In cylindrical co-ordinates

$$\nabla\times\mathbf{E} = \left(\frac{1}{\rho}\frac{\partial E_z}{\partial\phi} - \frac{\partial E_\phi}{\partial z}\right)\mathbf{a}_\rho + \left(\frac{\partial E_\rho}{\partial z} - \frac{\partial E_z}{\partial\rho}\right)\mathbf{a}_\phi$$
$$+ \frac{1}{\rho}\left(\frac{\rho\partial E_\phi}{\partial\rho} - \frac{\partial E_\rho}{\partial\phi}\right)\mathbf{a}_z \tag{5.94}$$

In this case,

$$\frac{\partial E_z}{\partial\phi} = \frac{\partial E_\phi}{\partial z} = \frac{\partial E_\rho}{\partial z} = \frac{\partial E_\phi}{\partial\rho} = \frac{\partial E_\rho}{\partial\phi} = 0 \tag{5.95}$$

Hence, for sine waves and putting $\rho = r$

$$curl\ \mathbf{E} = -\frac{\partial E_z}{\partial r}\mathbf{a}_\phi = -j\omega(B_\rho\mathbf{a}_\rho + B_\phi\mathbf{a}_\phi + B_z\mathbf{a}_z) = -j\omega B_\phi\mathbf{a}_\phi \tag{5.96}$$

This shows that B_ρ and B_z must be zero. Since the partial derivative may be replaced by a single variable derivative then

$$B_\phi = \frac{1}{j\omega} \frac{dE_z}{dr} \tag{5.97}$$

The electric field can be obtained from the applied current by $E_z = \rho J_z$ where ρ is the conductor resistivity, assumed independent of E. The current density, (5.38) in (5.105) is

$$J_z = J_a \frac{J_o(u)}{J_o(u_a)} \tag{5.98}$$

where J_a is the current density at the cylinder surface $r = a$ and $u_a = amj\sqrt{j}$. Substituting E_z for J_z in (5.97) gives

$$B_\phi = \frac{\rho}{j\omega} \frac{dJ_z}{dr} \tag{5.99}$$

$$B_\phi = j^{1/2} \frac{\rho m J_a}{\omega J_o(u_a)} \frac{dJ_o(u)}{du} \tag{5.100}$$

Using the Bessel identity $dJ_o(u)/du = -J_1(u)$, gives

$$B_\phi = -(j^{1/2}) \frac{\rho m J_a}{\omega} \frac{J_1(u)}{J_o(u_a)}, \quad r \le a. \tag{5.101}$$

or

$$B_\phi = \frac{\mu a J_a}{u_a} \frac{J_1(u)}{J_o(u_a)} \tag{5.102}$$

At low frequencies or $u \ll 1$, $J_o(u) = 1$ and $J_1(u) = u/2$, then (5.101) becomes

$$B_\phi = \frac{\mu I_o r}{2\pi a^2}, \quad r \le a. \tag{5.103}$$

This agrees with vector potential calculations for the dc case [20]. Generally, for $x < 1$, $J_1(x)/J_o(x) \approx x/2$ [91]. For the ac case, if $r = a$ and $u_a \le 1$, then (5.101) becomes

$$B_\phi = \mu J_a/2 = \frac{\mu I_o}{2\pi a} \tag{5.104}$$

5.6.2 Alternative expression for B_ϕ

J_a is given by (5.47)

$$J_a = \frac{I_o}{\pi a^2} \frac{u_a}{2} \frac{J_o(u_a)}{J_1(u_a)} \tag{5.105}$$

Substituting this into (5.102) gives

$$B_\phi = \frac{\mu I_o}{2\pi a} \frac{J_1(u)}{J_1(u_a)}, \quad r \le a. \tag{5.106}$$

This equation agrees with the expression given by J. J. Thomson's footnotes in Maxwell's book, after converting their units to SI [3] (p. 325).

5.6.3   *External magnetic field*

If we assume that the electric displacement current is negligible, the external magnetic field can be computed using Ampere's Law

$$\oint_l \mathbf{H}.d\mathbf{l} = I_o, \quad r \geq a. \tag{5.107}$$

Integrating Ampere's integral around the cylinder circumference with line element $dl = rd\phi$

$$\oint_l \mathbf{H}.d\mathbf{l} = \int_0^{2\pi} H_\phi rd\phi = H_\phi 2\pi r = I_o, \quad r \geq a. \tag{5.108}$$

Putting $B = \mu H$ gives the flux density outside the conductor

$$B_\phi = \frac{\mu I_o}{2\pi r}, \quad r \geq a. \tag{5.109}$$

agreeing with Biot–Savart calculations for the dc flux density at a perpendicular distance r from an infinite wire.

5.7   Magnetic field alternative solution

For the cylindrical conductor

$$curl\ \mathbf{H} = \left(\frac{1}{r}\frac{\partial H_z}{\partial \phi} - \frac{\partial H_\phi}{\partial z}\right)\mathbf{a_r} + \left(\frac{\partial H_r}{\partial z} - \frac{\partial H_z}{\partial r}\right)\mathbf{a_\phi}$$
$$+ \frac{1}{r}\left(\frac{\partial(rH_\phi)}{\partial r} - \frac{\partial H_r}{\partial \phi}\right)\mathbf{a_z} \tag{5.110}$$

For this case, with current only flowing in the axial direction – z, $H_z = H_r = 0$ and no variation of H with ϕ or z, but $H_\phi = f(r)$, then

$$curl\ \mathbf{H} = \frac{1}{r}\frac{\partial(rH_\phi)}{\partial r} = \frac{H_\phi}{r} + \frac{\partial H_\phi}{\partial r} \tag{5.111}$$

From (2.3) assuming no displacement current, then

$$curl\ \mathbf{H} = \mathbf{J} \tag{5.112}$$

and

$$J_z = \frac{H_\phi}{r} + \frac{\partial H_\phi}{\partial r} \tag{5.113}$$

Now since the electric field has only a z-component E_z, we have

$$curl\ \mathbf{E} = -\frac{\partial E_z}{\partial r}\mathbf{a_\phi} = -\rho\frac{\partial J_z}{\partial r}\mathbf{a_\phi} \tag{5.114}$$

From (5.113) and (5.114)

$$\frac{\partial J_z}{\partial r} = \frac{\partial}{\partial r}\left(\frac{H_\phi}{r}\right) + \frac{\partial^2 H_\phi}{\partial r^2} = \frac{1}{r}\frac{\partial H_\phi}{\partial r} - \frac{H_\phi}{r^2} + \frac{\partial^2 H_\phi}{\partial r^2} \tag{5.115}$$

Hence,

$$curl\ \mathbf{E} = \rho\frac{\partial J_z}{\partial r} = -\rho\left[\frac{1}{r}\frac{\partial H_\phi}{\partial r} - \frac{H_\phi}{r^2} + \frac{\partial^2 H_\phi}{\partial r^2}\right] \tag{5.116}$$

Also,

$$curl\ \mathbf{E} = -\mu\frac{\partial \mathbf{H}}{\partial t} \tag{5.117}$$

Thus,

$$\frac{\partial^2 H_\phi}{\partial r^2} + \frac{1}{r}\frac{\partial H_\phi}{\partial r} - \frac{H_\phi}{r^2} = \mu\sigma\frac{\partial \mathbf{H}_\phi}{\partial t} \tag{5.118}$$

which agrees with (5.13).

5.8 Graphical results for cylindrical conductors

The following graphical results were obtained using the MATLAB® programmes at the end of this chapter. The graphical results were plotted by copying the output text data to Easy Plot.

5.8.1 Current density

The current density was determined directly from (5.45) restated here as

$$J = \frac{j^{3/2}mI_o}{2\pi a}\frac{J_o(u)}{J_1(u_a)} = \frac{I_o}{\pi a^2}\frac{u_a}{2}\frac{J_o(u)}{J_1(u_a)} \tag{5.119}$$

As $\omega \rightarrow 0$, $J_o(u) \rightarrow 1$, $J_1(u_a) \rightarrow u_a/2$ and $J(r) \rightarrow \frac{I}{\pi a^2} = \bar{J}(r)$; the average current density. Figure 5.1 shows the current density for a copper wire radius $a = 1$ mm and $I = 1$A. Figure 5.1(a) shows the amplitude vs r/a plotted for three frequencies. At 1 khz the amplitude is constant across the whole diameter. At higher frequencies the current density decreases sharply from the surface. The current density at the surface, however, increases considerably with frequency, see in the following section. Figure 5.1(b) shows the current density complex components: real (J(R)), imaginary (J(I)) and modulus $|J|$. The sign of these components changes depending on r/a.

5.8.2 Surface current density

In Figure 5.2, the current density for which $r = a$ is plotted as a function of applied frequency. For frequencies which give a skin depth $\delta \geq a$, the surface current density

Wire current density from $J = j^{3/2}\,(mI/(2\pi a))J_0(u)/J_1(u_a)$

From Matlab wj1.txt
$I = 1$ A, $a = 1$mm

wj2.ep

Wire current density from $J = j^{3/2}\,(mI/(2\pi a))J_0(u)/J_1(u_a)$

From Matlab wj1.txt
$f = 100$ kHz, $I = 1$A, $a = 1$mm

wj1.ep

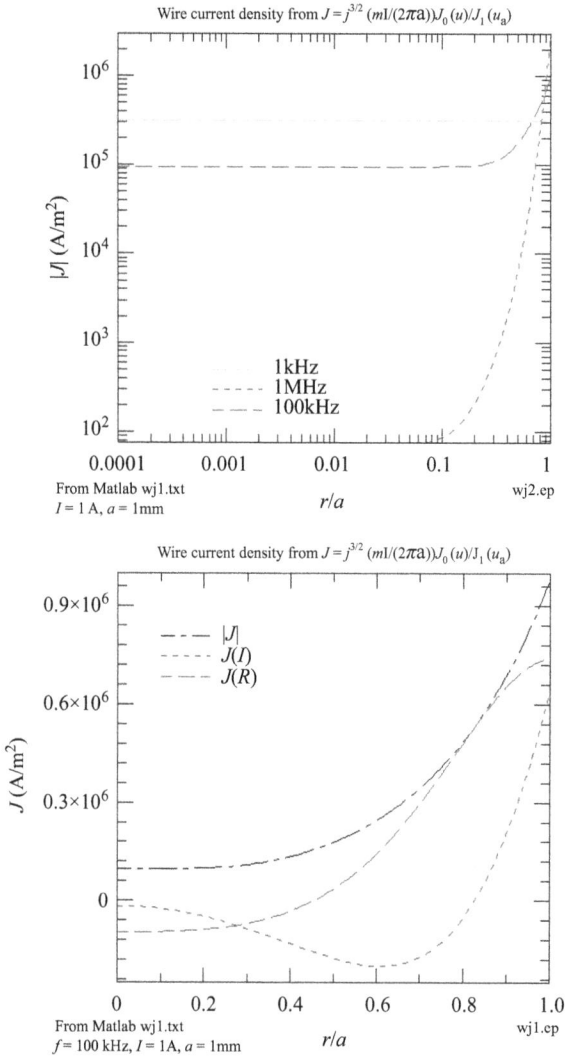

Figure 5.1   (a) Wire current density amplitude for three values of applied frequency from (5.45). (b) Wire current density complex components at 100 kHz plotted from (5.45).

is constant with $J_s = 318310\ A/m^2$. This corresponds to the average current density $J_{ave} = I/(\pi a^2) = 318310\ A/m^2$. For frequencies which give a skin depth $\delta \leq a$, the surface current density increases approximately as $f^{1/2}$ exceeding $100 J_{ave}$ at 1 GHz. This is because the area of conduction is reduced to a thin-walled tube with a sectional area $2\pi a\delta$. Hence, for $I = 1$ A then $J_s = 1/(2\pi \times 0.001\delta) = 159/\delta$. Such high current densities at the surface can melt the conductor.

Jaf.ep

Surface current density Ja

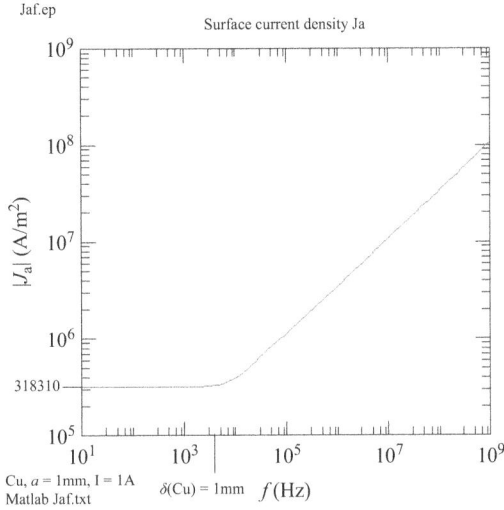

Figure 5.2 Surface current density $J_s = J_a$ when $r = a$

5.8.3 Impedance of conductor

The normalised impedance is obtained from (5.79) restated here as

$$Z_n = \frac{Z}{R_{dc}} = \frac{j^{3/2}a}{\sqrt{2}\delta} \frac{J_o(u)}{J_1(u_a)} = \frac{u_a}{2} \frac{J_o(u)}{J_1(u_a)} \tag{5.120}$$

As $\omega \to 0$, $Z \to R_{dc}$ as expected.

An approximate equation for the ac resistance was determined in the Skin Effect chapter (4.10) restated here as

$$\frac{R_{ac}}{R_{dc}} = \frac{1}{1 - (\frac{a-\delta}{a})^2} = \frac{1}{1 - (1 - 1/x)^2} \tag{5.121}$$

where $x = a/\delta$. Hence, if $\delta = a$ then $R_{ac} = R_{dc}$. If $\delta \ll a$ then $R_{ac} \gg R_{dc}$.

Figures 5.3 and 5.4 show the results of plotting (5.120) and (5.121) for a 1 mm radius copper wire. Figure 5.3 is a plot of normalised impedance vs $x = a/\delta$. It is seen that when the skin depth equals the conductor radius, that is, $a = \delta$, then the real and modulus components of the normalised impedance and the approximate ac resistance become equal to the dc resistance, that is, $|Z| = Z(I) = R_{ac} = R_{dc}$. As the skin depth becomes very large, the imaginary component of the normalised impedance tends to zero. This is because the normalised reactance is $X_n = Z_n(I) = X/R_{dc} = 0.5(a/\delta)^2$. Hence, as $\delta \gg a$ then $Z_n(I) \to 0$. If $\delta = a$ then $Z_n(I) = 1/2$.

It is also shown that for $a/\delta > 1$, the approximate equation for the ac resistance fits very well the real part of the impedance. If $a/\delta = x = 1$, a discontinuity occurs in R_{ac} due to the $1 - 1/x$ term in (5.121). Hence, this equation is not applicable for skin depths greater than the wire radius.

Figure 5.4 shows the frequency dependence of the absolute impedance and approximate ac resistance. The approximate equation also gives a good fit with $Z(R)$

Normalised impedance from eqn 6.33

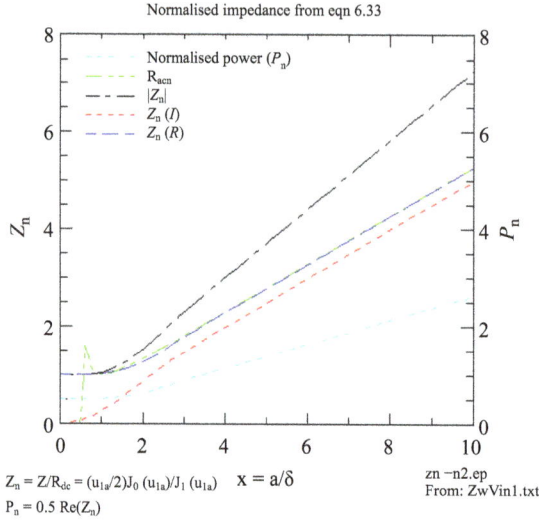

$Z_n = Z/R_{dc} = (u_{1a}/2)J_0 (u_{1a})/J_1 (u_{1a})$ $x = a/\delta$

$P_n = 0.5 \, \mathrm{Re}(Z_n)$

zn −n2.ep
From: ZwVin1.txt

Figure 5.3   Normalised impedance vs ratio of wire diameter to skin depth using (5.79) and (4.10). Also shown is the normalised power, P_n.

Impedance from eqn 6.33 and approx eqn 3.84

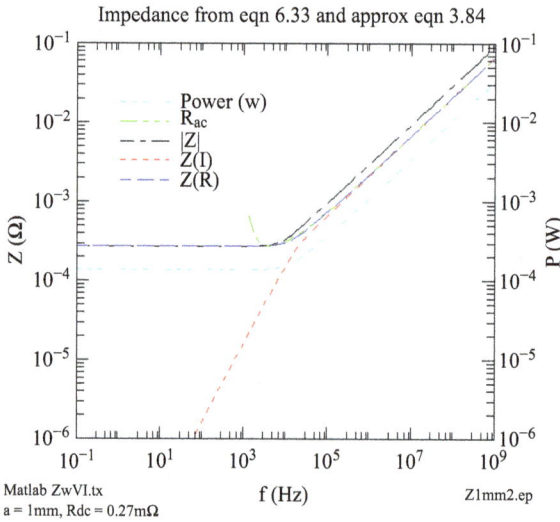

Matlab ZwVI.tx
a = 1mm, Rdc = 0.27mΩ

f (Hz)

Z1mm2.ep

Figure 5.4   As Figure 5.3, but absolute impedance vs frequency. Also shown is the power dissipated.

for frequencies above about 3.86 kHz. This corresponds to the frequency at which the skin depth equals the wire radius. For frequencies below 3.86 kHz, $|Z| = Z(R) = R_{dc} = 2.7 \times 10^{-4}$ Ω. As the frequency decreases towards zero and the imaginary impedance or reactance tends towards zero, that is, $Z(I) \to 0$.

5.8.4 Power dissipation

The power dissipated is given by (6.25).

$$P = <vi> = \frac{I_o^2}{2} Re[Z(\omega)] \tag{5.122}$$

where I_o is the applied current amplitude and the impedance is

$$Z = \frac{j^{3/2}l\rho m}{2\pi a} \frac{J_o(u)}{J_1(u_a)} = \frac{j^{3/2}R_{dc}a}{\sqrt{2}\delta} \frac{J_o(u)}{J_1(u_a)} \tag{5.123}$$

For a current of 1 A, the dissipation is then just half the real impedance as shown in Figure 5.4.

The normalised power dissipation is given by (6.28) restated here as

$$P_N = \frac{P}{I_o^2 R_{dc}} = j^{3/2} \frac{a}{2\delta} Re\left[\frac{J_o(u_a)}{J_1(u_a)}\right] = \frac{1}{2} Re\left[\frac{u_a J_o(u)}{2J_1(u_a)}\right] = \frac{1}{2} Re(Z_N) \tag{5.124}$$

For a current of 1 A, the dissipation is then just half the real normalised impedance as shown in Figure 5.3.

5.8.5 Magnetic flux density

The magnetic flux density, B, was derived for fields inside the wire, that is, $r \le a$, (5.101) repeated here as

$$B_\phi = -j^{1/2} \frac{\rho m J_a}{\omega} \frac{J_1(u)}{J_o(u_a)}, \quad r \le a. \tag{5.125}$$

At low frequencies and $u \ll 1$, $J_o(u) = 1$ and $J_1(u) = u/2$. Equation (5.101) becomes

$$B_\phi = -j^{1/2} \frac{\rho m J_a}{\omega} \frac{u}{2} = \frac{\mu I r}{2\pi a^2}, \quad u \ll 1 \tag{5.126}$$

where $J_a = I/\pi a^2$. This agrees with the dc calculation for the flux density inside the wire, [20]. Outside the wire it can be shown that the dc or low frequency magnetic field decreases as

$$B_\phi = \frac{\mu I}{2\pi r}, \quad r \ge a. \tag{5.127}$$

Figure 5.5 shows plots of the internal magnetic flux density (B) obtained from (5.101) produced by an ac current of 1 A flowing in a copper wire. (a) shows the complex field components, $|B|$, $B(R)$ and $B(I)$ plotted as a function of the ratio r/a and fixed frequency 1 *MHz*. (b) shows the magnitude only of the field plotted for three frequencies. It is clear from (a) that the magnetic flux density at 1 MHz is concentrated near the conductor surface decreasing significantly at $r/a = 0.6$ or 40 per cent penetration. At the other frequencies, the field decreases less rapidly but reduces to zero at the conductor centre in agreement with (5.125) and (5.126).

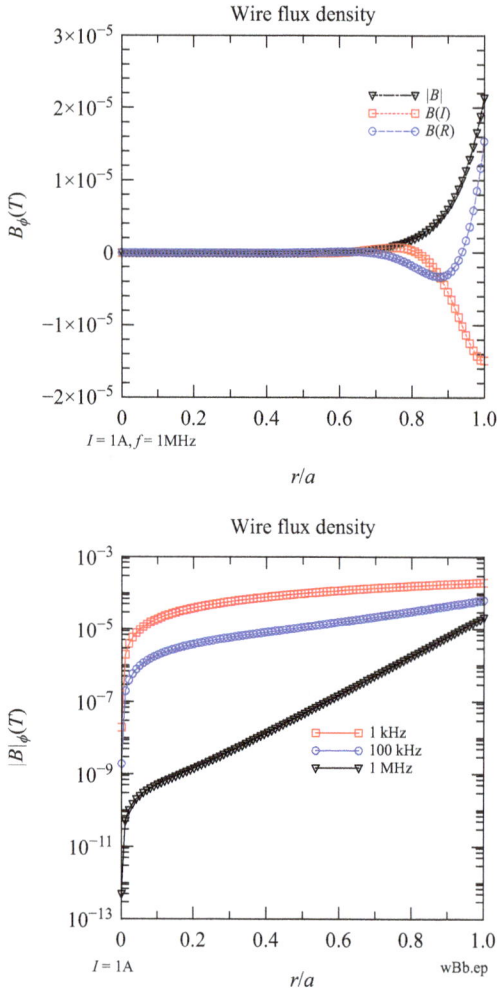

Figure 5.5 (a) *Complex field components,* $|B|$, $B(R)$ *and* $B(I)$ *plotted as a function of the ratio* r/a *and fixed frequency 1 MHz.(b) Magnitude only of the field* $|B|$ *plotted for three frequencies, 1100 and 1000 kHz.*

5.8.6 *MATLAB® programmes*

!wj1.txt MATLAB® analysis of wire current density.

format short e

for x = linspace(.0001,1,100)

muo=pi*4e-7;mur=1;mu=muo*mur;

rho=2.35e-8; !ohm m.

I=1; !Supply current amplitude 1A

f=1e3;w=2*pi*f;

a=0.001; r=a*x;!rod radius, m.

$del = (2 * rho/(w * mu))^{1/2}$; !skin depth

m=1.414 /del;$u = m * r * j * (j)^{1/2}; u1 = m * a * j * (j)^{1/2}$;

Jo=besselj(0,u);J0a=besselj(0,u1);J1a=besselj(1,u1);

$J = (j(3/2)) * (m * I/(2 * pi * a)) * Jo/J1a$; !Current density

JR=real(J);JI=imag(J);Jmag=abs(J);

disp([x Jmag])

end

Note: The ! symbol substitutes for the percentage symbol which is used for comments in MATLAB®.

%Zw VI.txt Matlab analysis of wire impedance equation from Z = V/I eqn 6.33.

```
format short e
%for x = linspace (.0001,.004,50)
for x = logspace (−1,9,100)

muo = pi*4e−7;
mur = 1;
mu = muo*mur;
rho = 1.7e−8;              %ohm m.
f = x;
%a = 1.27e−5;                        %radius, m. (50 SWG wire, diam = 0.001 in)
a = 0.001;
L = 5e−2;                  %Length of wire, m.
Rdc = rho*L/(pi*a^2);           %dc resistance
Ba = .001;                 %applied field (T)
w = 2*pi*f;
del = (2*rho/(w*mu))^.5;          %skin depth

m = 1.414 /del;

u = m*x*j*(j)^.5;
ul = m*a*j*(j)^.5;

J0 = besselj(0, ul);
J1 = besselj(1, ul);

Z = Rdc*(ul/2)*Jo/J1;

u2 = a/del;

Rac = Rdc/(1−((a−del)/a)^2);              %ac Resistance from approx cyl. eqn 3.84

ZR = real(Z);
ZI = imag(Z);
Zmag = abs(Z);

%is p([u2 ZR ZI Z mag Rac])
dis p([x ZR ZI Z mag Rac])
end
Rdc
```

Figure 5.6   MATLAB programme for impedance

%Bw .txt Matlab analysis of wire magnetic flux density.

```
format short e
for x = linspace(.0001,1,100)
%for x = logspace(,1,100)

muo = pi*4e−7;
mur = 1;
mu = muo*mur;
rho = 2.35 e−8;          %o hm m.
I = 1;                   %Applied current (A).
f = 1e5;
a = 0.001;              %rod radius, m.
r = a*x;
w = 2*pi*f;
del = (2*rho/(w*mu))^.5;     %skin depth
m = 1.414 /del;

u = m*r*j*(j)^.5;
ul = m*a*j*(j)^.5;
Ba = −(j^(1/2))*rho*m*I/(w*pi*a^2);

J1 = besselj(1,u);
Joa = besselj(0,ul);
B = Ba*J1/Joa;

BR = real(B);
BI = imag(B);
Bmag = abs(B);

%disp ([x BR BI Bmag])
disp ([x Bmag])
end
```

Figure 5.7   MATLAB programme for flux density

Section 5.8.6 provides the MATLAB programs used to calculate the wire current density, Figure 5.6 MATLAB program for the impedance and Figure 5.7 MATLAB program for flux density.

Chapter 6
Power dissipation in a cylindrical conductor

6.1 Introduction

This chapter on power determines the real part of the power dissipated in a *good conductor* and does not include the energy stored in the fields, power dissipated by dielectric loss or hysteresis loss, as discussed in the Power Flow chapter.

6.1.1 Power from Poynting theory

The time average power dissipation due to ohmic resistance alone (c.f. complex Poynting theory) is

$$P = \frac{\sigma}{2} \int_v \mathbf{E}.\mathbf{E}^* \, dv = \frac{\sigma}{2} \int_v \frac{\mathbf{J}}{\sigma}.\frac{\mathbf{J}^*}{\sigma} \, dv \tag{6.1}$$

The normal and conjugate current densities are

$$J(r,t) = J(r)e^{j\omega t}, \quad J^*(r,t) = J(r)e^{-j\omega t}, \quad J(r,t)J^*(r,t) = J^2(r) \tag{6.2}$$

Hence,

$$P = \frac{\rho}{2} \int_v \mathbf{J}(\mathbf{r}).\mathbf{J}(\mathbf{r})^* \, dv \tag{6.3}$$

For a cylindrical conductor, the solutions are

$$J(r) = C_1 J_o(u_1), \quad J(r)^* = C_2 J_o(u_2) \tag{6.4}$$

where $dv = 2\pi r dr l$ and

$$u_1 = mrj^{3/2}, \quad u_2 = mrj^{1/2} \tag{6.5}$$

Substitute in (6.3)

$$P = \frac{2\pi l\rho C_1 C_2}{2} \int_o^a r J_o(u_1) J_o(u_2) \, dr \tag{6.6}$$

From (A.41)

$$\int_o^a r J_o(u_1) J_o(u_2) \, dr = \frac{a}{2jm^2} \left[J_o(u_2)\frac{dJ_o(u_1)}{da} - J_o(u_1)\frac{dJ_o(u_2)}{da} \right] \tag{6.7}$$

Substitute for the Kelvin functions

$$\int_o^a rJ_o(u_1)J_o(u_2)\,dr = \frac{a}{2jm^2}[(ber_oma - jbei_oma)(ber'_oma + jbei'_oma)$$
$$- (ber_oma + jbei_oma)(ber'_oma - jbei'_oma)] \qquad (6.8)$$

where the prime ($'$) refers to differentiation with respect to a. Putting

$$a = ber_oma, \quad b = bei_oma, \quad a' = ber'_oma, \quad b = bei'_oma \qquad (6.9)$$

Hence,

$$\int_o^a rJ_o(u_1)J_o(u_2)\,dr = \frac{a}{2jm^2}[(a - jb)(a' + jb') - (a + jb)(a' - jb')] \qquad (6.10)$$

Multiplying out gives

$$\int_o^a rJ_o(u_1)J_o(u_2)\,dr = \frac{a}{m^2}(ab' - a'b) \qquad (6.11)$$

Substituting into (6.7)

$$P = \frac{2\pi l\rho C_1 C_2}{2}\frac{a}{m^2}(ber_oma\,bei'_oma - ber'_oma\,bei_oma) \qquad (6.12)$$

From (5.62), we obtained

$$I_{o1} = \frac{2\pi C_1}{m}[bei'_o(ma) - jber'_o(ma)] \qquad (6.13)$$

We need to use a similar method to determine I_{o2} and C_2. From these, we can obtain $C_1 C_2$ from $I_{o1}I_{o2}$.

$$I_{o2} = \int_s \mathbf{J}^*.d\mathbf{s} = C_2 \int_0^a J_o(u_2)2\pi r dr \qquad (6.14)$$

$$I_{o2} = \frac{2\pi C_2}{(mj^{1/2})^2}\int_0^{u_{2a}} u_2 J_o(u_2)\,du_2 \qquad (6.15)$$

Using the identity $\int uJ_o(u)du = uJ_1(u)$

$$I_{o2} = \frac{2\pi C_2}{(mj^{1/2})^2}u_{2a}J_1(u_{2a}) = \frac{2\pi C_2 a}{mj^{1/2}}J_1(u_{2a}) \qquad (6.16)$$

Substitute for the Kelvin function

$$I_{o2} = \frac{2\pi C_2 a}{mj^{1/2}}j^{-1/2}(ber'_oma - jbei'_oma) \qquad (6.17)$$

Hence,

$$I_{o2} = -\frac{2\pi C_2 a}{m}(bei'_oma + jber'_oma) \qquad (6.18)$$

The amplitude of the supply current is then

$$I_o^2 = I_{o1}I_{o2} = \left(\frac{2\pi a}{m}\right)^2 C_1 C_2(ber'^2_oma + bei'^2_oma) \qquad (6.19)$$

Hence,

$$C_1 C_2 = \left(\frac{m}{2\pi a}\right)^2 \frac{I_o^2}{ber_o'^2 ma + bei_o'^2 ma} \tag{6.20}$$

The power dissipation (6.12) then becomes

$$P = \left(\frac{2\pi l \rho a}{2m^2}\right) \left(\frac{m}{2\pi a}\right)^2 I_o^2 m \left(\frac{ber_o ma \, bei_o' ma - ber_o' ma \, bei_o ma}{ber_o'^2 ma + bei_o'^2 ma}\right) \tag{6.21}$$

where the differentiation is with respect to ma, that is, $m\frac{d}{dma}$. Hence, the extra m in the numerator. The above equation may now be simplified to

$$P = \frac{I_o^2 R_{dc}}{2\sqrt{2}} \frac{a}{\delta} \left(\frac{ber_o ma \, bei_o' ma - ber_o' ma \, bei_o ma}{ber_o'^2 ma + bei_o'^2 ma}\right) \tag{6.22}$$

where $R_{dc} = \frac{\rho l}{\pi a^2}$ and $m = \sqrt{2}/\delta$.

If we put $P = I_o^2 R_{ac}$ where R_{ac} is the *ac* resistance, then

$$\frac{R_{ac}}{R_{dc}} = \frac{a}{2\sqrt{2}\delta} \left(\frac{ber_o ma \, bei_o' ma - ber_o' ma \, bei_o ma}{ber_o'^2 ma + bei_o'^2 ma}\right) \tag{6.23}$$

The normalised *ac* power dissipation is the ratio of the *ac* dissipation to the *dc* dissipation

$$P_N = \frac{P}{|I|^2 R_{dc}} = \frac{R_{ac}}{R_{dc}} \tag{6.24}$$

Thus it is identical to the real part of the normalised *ac* resistance.

6.1.2 Power from impedance

The power dissipated may be determined from the impedance calculated here using the time average power (3.126)

$$P = <vi> = \frac{I_o^2}{2} Re[Z(\omega)] \tag{6.25}$$

where $\hat{I}\hat{I}^* = I_o^2$ is the applied current amplitude and the impedance is from (5.78)

$$Z = \frac{j^{3/2} l \rho m}{2\pi a} \frac{J_o(u)}{J_1(u_a)} = \frac{j^{3/2} R_{dc} a}{\sqrt{2}\delta} \frac{J_o(u)}{J_1(u_a)} \tag{6.26}$$

Substituting for the real part of the impedance (5.89) gives

$$P = \frac{I_o^2 R_{dc} a}{2\sqrt{2}\delta} \left[\frac{ber_o(mr)bei_o'(ma) - bei_o(mr)ber_o'(ma)}{ber_o'^2(ma) + bei_o'^2(ma)}\right] \tag{6.27}$$

which agrees with the previous calculation (6.22).

The normalised power dissipation is then

$$P_N = \frac{P}{I_o^2 R_{dc}} = j^{3/2} \frac{a}{2\delta} Re\left[\frac{J_o(u_a)}{J_1(u_a)}\right] = \frac{1}{2} Re\left[\frac{u_a J_o(u)}{2J_1(u_a)}\right] \tag{6.28}$$

These are useful for calculations using MATLAB® or other software. We may also use (5.45) to give

$$P = \frac{I_o^2 R_{dc}}{2} Re \left[\frac{J(r)}{\bar{J}(r)} \right] \tag{6.29}$$

where $J(r)$ is the radial dependent current density. $\bar{J}(r) = I_o/\pi a^2$ is the average current density, which is equal to the dc current density. Hence,

$$P = \frac{I_o \rho l}{2} Re[J(r)] = \frac{I_o}{2} Re[V(r)] \tag{6.30}$$

where $V(r) = l\rho J(r)$. The time-averaged power dissipated is thus half the IV product. This is only of general interest since the Bessel function dependent current density $J(r)$ or voltage $V(r)$ have to be determined. However, the result supports the validity of the analysis.

Chapter 7

Inductance and resistance of cylindrical conductors – analysis and experimental measurements

7.1   Introduction

This chapter mainly concerns the analysis and experimental measurements of the inductance and resistance of cylindrical conductors. The theory of the internal resistance of metals as a function of frequency leading to the skin effect has been widely reported over many years, ever since James Clerk Maxwell's seminal work, 'Treatise on Electricity and Magnetism' [3,38,45]. Excellent accounts have been presented concerning the internal impedance of copper conductors including details of the ac resistance and internal inductance [88]. Recent work has largely concentrated on the accuracy of the theories and Bessel functions used in the analysis particularly at high frequencies for application in rf devices [89]. The skin effect in ferrous metals and in particular railway tracks have also received attention over the years including analysis of the internal inductance [90].

Experimental measurements of the skin effect, particularly the internal inductance (L_i) in non-ferrous conductors have been much less investigated. This may be because L_i is generally considerably less than the external inductance L_e, which depends on the geometry of the conductor being studied. Attempts were made previously to measure L_i in copper conductors but with varying success [91]. At low frequencies or d.c. L_i of non-ferrous metals is constant with maximum value of $50\,\mu_r$ nHm^{-1} where μ_r is the relative magnetic permeability. For copper $\mu_r \approx 1$ giving $L_i = 50$ nHm^{-1}. However, in ferrous metals μ_r may be very large and L_i more significant. At frequencies above a certain threshold, L_i decreases with frequency according to $f^{-1/2}$ and the resistance increases with frequency according to $f^{+1/2}$. At low frequencies close to d.c., L_i in cylindrical conductors is independent of conductor radius. However, the *threshold frequency* is dependent on the conductor radius (see Section 7.6).

In this chapter, we consider a cylindrical conductor with radius a and length l where the current flows in the axial direction. The applied current is assumed to be either steady state (dc) or sinusoidal (ac). We first briefly review the steady state, d.c., calculation of the internal and external inductance of single wires, two-wire and coaxial transmission lines. This is followed by an analysis of the internal inductance for sinusoidal currents determined by equating energies in the inductance with the energy in the magnetic fields. To achieve this, equations for the sinusoidal internal magnetic field are derived in detail.

The ohmic loss and internal ac resistance are then determined from power dissipation in the conductor. For the case of the internal inductance and resistance, the low frequency limit is obtained and compared with the steady state calculations.

Experimental measurements of the internal inductance and resistance of some non-ferrous and ferrous metals are then presented, followed by a discussion and summary.

7.2 Inductance

The term 'inductance' arises from Michael Faraday's experiments on electromagnetic induction [3]. In one experiment, he wound copper wire around opposite sides of an iron ring (modern-day toroidal transformer). He noticed that when the current was interrupted in one coil, it induced a voltage in the other coil. From these experiments, he produced the famous Faraday law of electromagnetic induction: 'The emf around a closed path is equal to the negative time rate of change of the magnetic flux enclosed by the path'. Although Faraday avoided the use of mathematical terminology, preferring to describe his observations in terms of the written word, today we use the mathematical form

$$V_{emf} = -\frac{d\Phi}{dt} \tag{7.1}$$

where V_{emf} is the electromagnetic force around the closed path and Φ the magnetic flux enclosed by the path. Since the magnetic flux is proportional to the current in the wire (I), we can write as follows:

$$\Phi = LI, \ \ or \ \ L = \frac{\Phi}{I} \tag{7.2}$$

where L is a constant associated with electromagnetic induction and referred to as the magnetic *inductance*, defined by the ratio of the magnetic flux to the current in the wire, also associated with this is the case of mutual inductance due to magnetic coupling with external components. In the following, we only consider the case of inductance of the wire itself, i.e. *self-inductance*. Equations for the self-inductance may be obtained for the steady state d.c. case and the time-dependent a.c. case.

7.2.1 Steady state (dc) calculations of self-inductance

In this case, the direct current I flows uniformly through the cross section of the conductor. We identify two cases: *internal inductance* inside the wire, $r \leq a$ where a is the radius of the wire and *external inductance* outside the wire $r \geq a$.

Internal self-inductance
For this case, the magnetic flux density is [43]

$$B_\phi = \frac{\mu I r}{2\pi a^2} \tag{7.3}$$

where $\mu = \mu_o \mu_r$. μ_o is the permeability of free space $(4\pi \times 10^{-7} \text{ H/m})$. μ_r is the relative permeability of the conductor. In this case, the current through the area πr^2 is only a fraction of the total area πa^2. Thus, we can write as follows:

$$I_r = I\frac{r^2}{a^2} \tag{7.4}$$

The internal magnetic flux through the rotational area $ds = ldr$ is then

$$\Psi_i = \int_s B_\phi ds = l \int_0^a B_\phi dr = \frac{l\mu I}{2\pi a^2} \int_0^a r\frac{r^2}{a^2} dr = \frac{l\mu I}{2\pi a^2}\frac{a^4}{4a^2} = \frac{l\mu I}{8\pi} \tag{7.5}$$

Hence the internal inductance $L_i = \Psi/I$ is

$$L_i = \frac{l\mu}{8\pi} = 50\mu_r \text{ nHm}^{-1} \tag{7.6}$$

The internal inductance is therefore only dependent on the magnetic permeability μ_r and independent of the wire's radius. However, the change in L_i with frequency does depend on conductor radius (see later sections).

External self-inductance
For a single solid cylindrical conductor or wire the determination of the external self inductance, $r \geq a$, is more complex and is given by [45,74]

$$L_e = \frac{\mu_o l}{2\pi}\left[\ln\frac{2l}{a} - 1\right] \tag{7.7}$$

The total inductance for a single solid cylindrical conductor or wire is then the sum of the internal inductance (Li) and external inductance (Le). Hence,

$$L = \frac{\mu_o l}{2\pi}\left[\ln\frac{2l}{a} - 1\right] + \frac{l\mu_o\mu_r}{8\pi} \tag{7.8}$$

Twin-wire Transmission Line
For a twin-wire transmission line calculations using vector potentials [20], the total inductance of the two single lines, each l metres long and spaced 2b apart, is

$$L_{TL} = \frac{l\mu_o}{\pi}\ln\frac{2b}{a} + \frac{l\mu_o\mu_r}{4\pi} \tag{7.9}$$

Note that for the twin-wire transmission line case, only the external inductance is affected by the distance between the conductors, 2b. If the distance between the conductors is close to the conductor radius, that is, $2b \approx a$, the external inductance is no longer given by (7.7), but becomes [38,92]

$$L_e = \frac{\mu_o}{\pi}\cosh^{-1}\frac{b}{a} \tag{7.10}$$

Coaxial Transmission Line
For a coaxial transmission line core radius a, screen radius b, the external magnetic field is

$$B_\phi = \frac{\mu_o I}{2\pi r} \tag{7.11}$$

$$B_\phi = \int_s B_\phi ds = l \int_a^b B_\phi dr = \frac{l\mu_o I}{2\pi} \int_a^b \frac{dr}{r} = \frac{l\mu_o I}{2\pi} \ln \frac{b}{a} \qquad (7.12)$$

Hence, the external inductance $L_e = \Psi_e/I$ is

$$L_{ce} = \frac{l\mu_o}{2\pi} \ln \frac{b}{a} \qquad (7.13)$$

The sum of the internal inductance and external inductance is then

$$L = \frac{l\mu_o}{2\pi} \ln \frac{b}{a} + \frac{l\mu_o \mu_r}{8\pi} \qquad (7.14)$$

7.2.2   Internal inductance-energy method

The steady state energy density in a magnetic field is [20]

$$w_M = \frac{1}{2}\mu H^2, \quad Jm^{-3} \qquad (7.15)$$

This can also be expressed in terms of current density and vector magnetic potential

$$W_M = \frac{1}{2} \int_V \mathbf{J}.\mathbf{A} \, dv \qquad (7.16)$$

where $\mathbf{J}$ is the current density and $\mathbf{A}$ is the vector potential. Since both of these parameters are proportional to current I, we can write $\mathbf{J}.\mathbf{A} = L_1 I^2$ where L_1 the constant of proportionality or inductance per m^3. In terms of energy density, this becomes $w_M = \frac{1}{2}L_1 I^2$. Equating this with (7.15) gives

$$\frac{1}{2}L_1 I^2 = \frac{1}{2}\mu H^2, \quad Jm^{-3}, \quad L_1 = \frac{1}{I^2}\mu H^2, \quad Hm^{-3} \qquad (7.17)$$

The inductance in Henries is $L = \int_V L_1 dv$. Hence

$$L = \frac{1}{I^2} \int_V \mu H^2 dv, \quad H \qquad (7.18)$$

For $r \leq a$ the internal magnetic field is

$$H_\phi = \frac{Ir}{2\pi a^2} \qquad (7.19)$$

The internal energy with relative permeability μ_r and inductance L_i are

$$\frac{1}{2}L_i I^2 = \frac{\mu_o \mu_r}{2} \int_0^a \left[\frac{Ir}{2\pi a^2}\right]^2 2\pi r dr l \qquad (7.20)$$

The internal inductance is then

$$L_i = \frac{\mu_o \mu_r l}{2\pi a^4} \int_0^a r^3 dr = \frac{\mu_o \mu_r l}{2\pi a^4} \frac{a^4}{4} \qquad (7.21)$$

$$L_i = \frac{\mu_o \mu_r l}{8\pi} \qquad (7.22)$$

which agrees with (7.6).

7.3 Sinusoidal fields

The time average energy relation is

$$\frac{1}{2}L<I^2> = \frac{1}{4}\int_V \mu\mathbf{H}_\phi.\mathbf{H}_\phi^* \, dV \tag{7.23}$$

where $\mathbf{H}_\phi^*$ is the complex conjugate of $\mathbf{H}_\phi$. The time average $<I^2> = I_0^2/2$ where I_0 is the current amplitude. The internal inductance is then

$$L_i = \frac{\mu}{I_0^2}\int_V \mathbf{H}_\phi.\mathbf{H}_\phi^* \, dV \tag{7.24}$$

7.3.1 Internal magnetic field

The magnetic field is obtained from Maxwell's equation

$$curl \, \mathbf{E} = -\frac{\partial \mathbf{B}}{\partial t} \tag{7.25}$$

which for cylindrical co-ordinates and sinusoidal fields

$$curl \, \mathbf{E} = -\frac{\partial E_z}{\partial r}\mathbf{a}_\phi = -j\omega B_\phi\mathbf{a}_\phi \tag{7.26}$$

Hence, since B_ϕ and E_z are functions of r only and $E_z = \rho J_z$, we can write

$$B_\phi(r) = (1/j\omega)\frac{dE_z(r)}{dr} = (\rho/j\omega)\frac{dJ_z(r)}{dr} \tag{7.27}$$

$$B_\phi = -j^{1/2}\frac{\rho m J_a}{\omega}\frac{J_1(u)}{J_0(u_a)}, \quad r \le a. \tag{7.28}$$

or

$$B_\phi = -\frac{\rho J_a}{\omega a}u_{2a}\frac{J_1(u_1)}{J_0(u_{1a})} \tag{7.29}$$

and it's conjugate

$$B_\phi^* = \frac{\rho J_a}{\omega a}u_{1a}\frac{J_1(u_2)}{J_0(u_{2a})} \tag{7.30}$$

where J_a is the current density at $r = a$. J_o and J_1 are zero and first-order Bessel functions of the first kind, respectively, with arguments

$$u_1 = mrj^{3/2}, \; u_{1a} = maj^{3/2}, u_2 = mrj^{1/2}, \; u_{2a} = maj^{1/2}, m^2 = \omega\mu\sigma. \tag{7.31}$$

We assume that the permeability μ and conductivity σ are constants. The internal magnetic field given by (7.28) is plotted in Figure 7.1 for a copper wire with the specified dimensions.

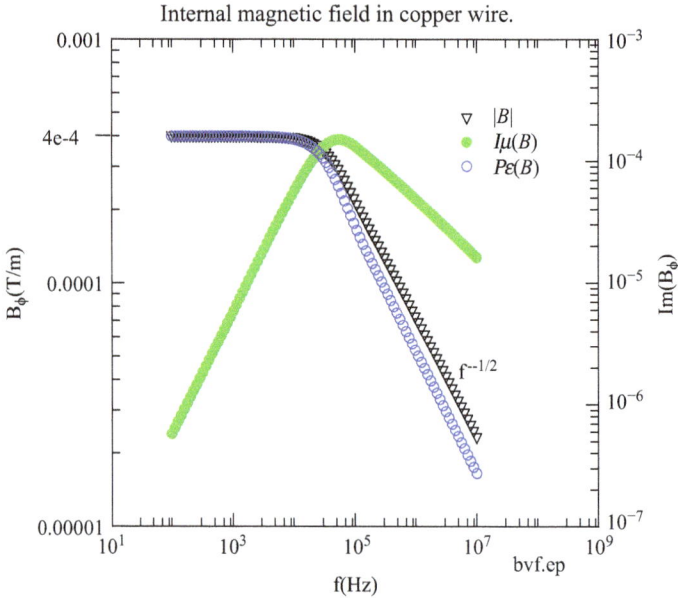

*Figure 7.1   Internal magnetic field in copper wire, radius 0.5 mm, length 1.0 m and
I = 1.0 A. Plotted from (7.28)*

7.3.2   Internal energy and inductance

The time average energy in the magnetic field is

$$W_M = \frac{2\pi}{4\mu} \int_0^a B_\phi B_\phi^* r dr \qquad (7.32)$$

Substitute for (7.29) and (7.30)

$$W_M = \frac{2\pi}{4\mu} B_a B_a^* \int_0^a J_1(u_1) J_1(u_2) r dr \qquad (7.33)$$

where

$$B_a = -\frac{\rho J_a}{\omega a} \frac{u_{2a}}{J_o(u_{1a})} \qquad (7.34)$$

$$B_a^* = \frac{\rho J_a}{\omega a} \frac{u_{1a}}{J_o(u_{2a})} \qquad (7.35)$$

Equating the inductive energy and the magnetic energy

$$\frac{1}{2} L < I^2 >= W_M \qquad (7.36)$$

The time average $< I^2 >= I_{rms}^2/2$ where I_{rms} is the rms current amplitude. The
internal inductance is then

$$L_i = 2W_M / I_{rms}^2 \qquad (7.37)$$

$$L_i = \frac{\pi}{\mu I_{rms}^2} B_a B_a^* \int_0^a J_1(u_1)J_1(u_2)rdr \tag{7.38}$$

The integral is given by [76]

$$\int_0^a xJ_1(\alpha x)J_1(\beta x)dx = \frac{x}{\alpha^2 - \beta^2}[\alpha J_1(\alpha x)J_o(\beta x) - \alpha J_o(\alpha x)J_1(\beta x)] \tag{7.39}$$

Putting

$$u_1 = mrj^{3/2} = \alpha x, \quad u_{1a} = maj^{3/2}\alpha a, \quad u_2 = mrj^{1/2} = \beta x,$$
$$u_{2a} = maj^{1/2} = \beta a. \tag{7.40}$$

Hence,

$$\int_0^a rJ_1(u_1)J_1(u_2)dr = \frac{a^2}{u_{1a}^2 - u_{2a}^2}[u_{2a}J_1(u_{1a})J_o(u_{2a}) - u_{1a}J_o(u_{1a})J_1(u_{2a})] \tag{7.41}$$

7.4   Current amplitude

In terms of the magnetic field the total current is given by the Biot–Savart law [20]

$$I = \frac{2\pi a B_\phi}{\mu} \tag{7.42}$$

or

$$I = -j^{1/2}\frac{2\pi a\rho mJ_a}{\mu\omega}\frac{J_1(u_1)}{J_o(u_{1a})} = -j^{1/2}\frac{2\pi aJ_a}{m}\frac{J_1(u_1)}{J_o(u_{1a})} \tag{7.43}$$

The complex conjugate is

$$I^* = j^{1/2}\frac{2\pi aJ_a}{m}\frac{J_1(u_2)}{J_o(u_{2a})} \tag{7.44}$$

$I_{rms} = I/\sqrt{2}$, $I_{rms}^2 = I^2/2$. Hence,

$$I_{rms}^2 = 2\left[\frac{2\pi aJ_a}{m}\right]^2 \frac{J_1(u_1)J_1(u_{1a})}{J_0(u_2)J_o(u_{2a})} \tag{7.45}$$

7.4.1   Low-frequency approximation

At low frequencies or $u \ll 1$, $J_o(u) = 1$ and $J_1(u) = u/2$, then

$$\int_0^a rJ_1(u_1)J_1(u_2)dr = -\frac{m^2}{4}\frac{a^4}{4} \tag{7.46}$$

$$I_{rms}^2 = -\frac{1}{2}\pi^2 a^4 J_a^2 \tag{7.47}$$

$$B_a B_a^* = \left[\frac{\mu J_a}{m}\right]^2 \tag{7.48}$$

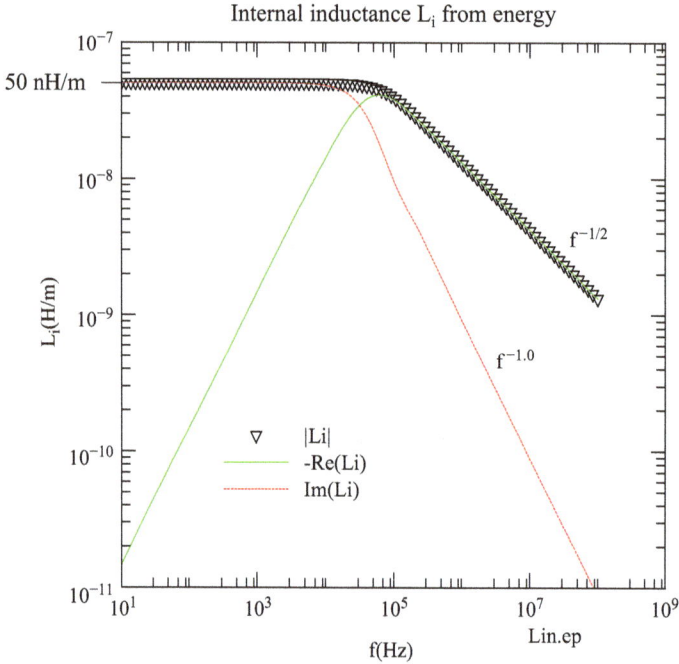

Figure 7.2   Internal inductance L_i for copper wire, radius 0.5 mm, length 1.0 m, $I = 1.0$ A from (7.38) to (7.47)

Substitution of the last three equations into (7.38) gives the LF internal inductance

$$L_{io} = \frac{\mu}{8\pi} \; Hm^{-1} = 50\mu_r \; nHm^{-1} \tag{7.49}$$

This is the same as that obtained for the steady state calculation (7.6).

Figure 7.2 shows a plot of the inductance L_i from (7.38) to (7.47) and associated equations using MATLAB®.

7.5   Ohmic loss and a.c. resistance

The power dissipation density due to ohmic resistance alone is

$$p = \mathbf{E.J} \; Jm^{-3} \tag{7.50}$$

Assuming an ohmic conductor with $J = \sigma E$ then the time average dissipation in volume V is

$$<P> = (\sigma/2) \int_V \mathbf{E.E}^* dV = (\rho/2) \int_V \mathbf{J.J}^* dV \tag{7.51}$$

For a solid cylinder of unit length and radius a,

$$< P > = (\rho/2) \int_0^a J_z J_z^* 2\pi r dr \tag{7.52}$$

In this case the current density is given by (5.98) in Appendix

$$J_z = J_a \frac{J_o(u_1)}{J_o(u_{1a})}, \quad J_z^* = J_a \frac{J_o(u_2)}{J_o(u_{2a})} \tag{7.53}$$

where J_z^* is the complex conjugate of J_z, J_a is the current density at the cylinder surface $r = a$ and the Bessel arguments u are given in (7.40). Substituting these two equations into (7.52) gives

$$< P > = \frac{\pi \rho J_a^2}{J_o(u_{1a})J_o(u_{2a})} \int_0^a J_o(u_1)J_o(u_2) r dr \tag{7.54}$$

From Poynting's theorem the time average power is also given by

$$< P > = \frac{I_o^2}{2} Re[Z(\omega)] = I_{rms}^2 Re[Z(\omega)] \tag{7.55}$$

where I_o is the current amplitude and $Re[Z(\omega)]$ is the real part of the frequency dependent complex impedance. Hence,

$$Re[Z(\omega)] = \frac{<P>}{I_{rms}^2} = \frac{\pi \rho J_a^2}{I_{rms}^2 J_o(u_{1a})J_o(u_{2a})} \int_0^a J_o(u_1)J_o(u_2) r dr \tag{7.56}$$

The *rms* current is given by (7.45). Substituting for this leads to

$$Re[Z(\omega)] = \frac{R_{dc} m^2}{2J_1(u_{1a})J_1(u_{2a})} \int_0^a J_o(u_1)J_o(u_2) r dr \quad \Omega/m \tag{7.57}$$

where $m^2 = \sigma \mu \omega$ and steady state dc resistance $R_{dc} = \rho/(\pi a^2)$. A dimensions check gives: $m^2 = \sigma \mu \omega = 1/metres^2$, the integral $= m^2$, denominator is dimensionless and $R_{dc} = \Omega/m$.

At low frequencies or $u \ll 1$, $J_o(u) = 1$ and $J_1(u) = u/2$ then

$$Re[Z(\omega)] = \frac{R_{dc} m^2}{2} \frac{a^2/2}{(u_{1a}/2)(u_{2a}/2)} = \frac{R_{dc}(ma)^2}{4} \frac{4}{-(ma)^2} = -R_{dc} \tag{7.58}$$

Hence, $|Re[Z(\omega)]| = R_{dc}$ at low frequencies as expected.

Figure 7.3 shows plots of the internal energy (W), inductance (L_i) and ac resistance (R_{ac}) (7.33), (7.37), (7.57), respectively.

7.6   Frequency response of internal inductance and resistance

The internal impedance of a cylindrical conductor is given by (5.79)

$$Z_i = R_{dc} \frac{u_a}{2} \frac{J_o(u_a)}{J_1(u_a)} \tag{7.59}$$

Copper wire Internal inductance L_i, resistance R_{ac} and energy W.

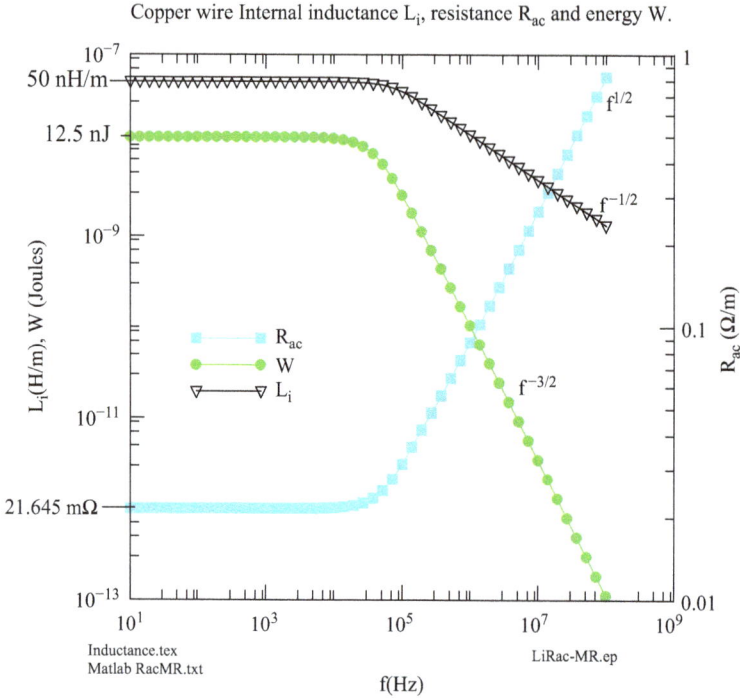

Figure 7.3   Internal inductance (L_i), resistance (R_{ac}) and energy (W). Copper wire, radius 0.5 mm, length 1.0 m and I = 1.0 A.

where R_{dc} is the steady state d.c. resistance of the conductor, $J_o(u_a)$ and $J_1(u_a)$ are zero-order and first-order Bessel functions of the first kind, respectively, $u_a = j^{3/2}\sqrt{2}a/\delta$, a is the wire radius, $\delta = \sqrt{2\rho/\omega\mu}$ is the skin depth of wire resistivity ρ, permeability $\mu = \mu_o\mu_r$ and angular frequency ω.

The internal resistance and internal inductance are then given by the real and imaginary parts of the complex impedance

$$R_i = Re(Z_i), \ L_i = \frac{Im(Z_i)}{\omega} \tag{7.60}$$

In order to obtain low and high frequency approximations we require the series expansion of (7.59). This is given by

$$\frac{u_a}{2}\frac{J_o(u_a)}{J_1(u_a)} = 1 - \frac{u_a^2}{8} - \frac{u_a^4}{192} - \frac{u_a^6}{3072} - \frac{u_a^8}{46080} - \cdots \tag{7.61}$$

which is proved in [95,104].

7.6.1 Low-frequency approximation

Taking the first three terms only of (7.61) and substituting into (7.59) gives

$$\mathbf{Z_i} = R_{dc} \left[1 + \frac{j}{4} \left(\frac{a}{\delta}\right)^2 + \frac{1}{48} \left(\frac{a}{\delta}\right)^4 \right] \tag{7.62}$$

Hence, the low-frequency resistance is

$$Re(Z_i) = R_i = R_{dc} \left[1 + \frac{1}{48} \left(\frac{a}{\delta}\right)^4 \right] \tag{7.63}$$

which agrees with Landau *et al.* [12]. This may be re-written as

$$R_i = R_{dc} + \frac{l^2 \mu^2 f^2}{48 R_{dc}} \tag{7.64}$$

The internal reactance is given by the imaginary part of (7.62)

$$Im(Z_i) = R_{dc} \frac{1}{4} \left(\frac{a}{\delta}\right)^2 \tag{7.65}$$

and the internal inductance is

$$L_{int} = \frac{Im(Z_{int})}{\omega} = \frac{R_{dc}}{4\omega} \left(\frac{a}{\delta}\right)^2 \tag{7.66}$$

Substituting for $R_{dc} = \rho l/(\pi a^2)$ and the skin depth equation, yields

$$L_{int} = \frac{l\mu}{8\pi} = 50\mu_r \ nH/m \tag{7.67}$$

which is the same as that obtained from steady state calculations [20]. Thus, in this low-frequency approximation, the ac resistance increases as f^2 and the inductance is constant. The approximate low frequency (7.63) agrees with the full Bessel function analysis (7.59) for values of $a/\delta \leq 1$, but diverges for higher values of a/δ.

7.6.2 High-frequency approximation

For high values of the Bessel function arguments $u_a \gg 1$ corresponding to high frequencies, then $J_0(u_a)/J_1(u_a) = -j$ [38]. Hence (7.59) becomes

$$\mathbf{Z_i} = R_{dc} \frac{u_a}{2} (-j) \tag{7.68}$$

and this leads to

$$\mathbf{Z_i} = \frac{R_{dc}}{2} (a/\delta)(1+j) = \frac{l}{2\pi a} (\rho/\delta)(1+j) \tag{7.69}$$

Thus,

$$R_i = \frac{R_{dc}}{2} (a/\delta) = \frac{aR_{dc}}{2} f^{1/2} \sqrt{\pi\mu/\rho} \tag{7.70}$$

Hence, at high frequencies, the real part of the impedance or the ac resistance is proportional to the square root of the applied frequency. Also from (7.69)

$$R_i = X_i = \omega L_i \tag{7.71}$$

Thus,

$$L_i = \frac{R_{dc}}{2\omega}(a/\delta) = \frac{l}{2\pi a}\left(\frac{\mu\rho}{2\omega}\right)^{1/2} \tag{7.72}$$

or

$$L_i = kf^{-1/2}, \quad R_i = 2\pi kf^{1/2} \tag{7.73}$$

where

$$k = \frac{l}{4\pi a}\left(\frac{\mu\rho}{\pi}\right)^{1/2} \tag{7.74}$$

Figure 7.4 shows R_i and L_i frequency response for Cu wire from (7.59) using Bessel functions. The Cu wire was 1 mm diameter, 1 m long, relative permeability $\mu_r = 1$, resistivity 17 $n\Omega m^{-1}$ giving $R_{dc} = 21.645$ mΩ. Equation (7.74) gives $k = 1.3124 \times 10^{-5}$.

The sloping straight lines in the figure follow (7.73) intercepting with the horizontal constant straight lines at a frequency $f = f_c$. This occurs for

$$R_i = R_{io} = R_{dc}, \quad L_i = L_{io} = 50\mu_r \ nH/m. \tag{7.75}$$

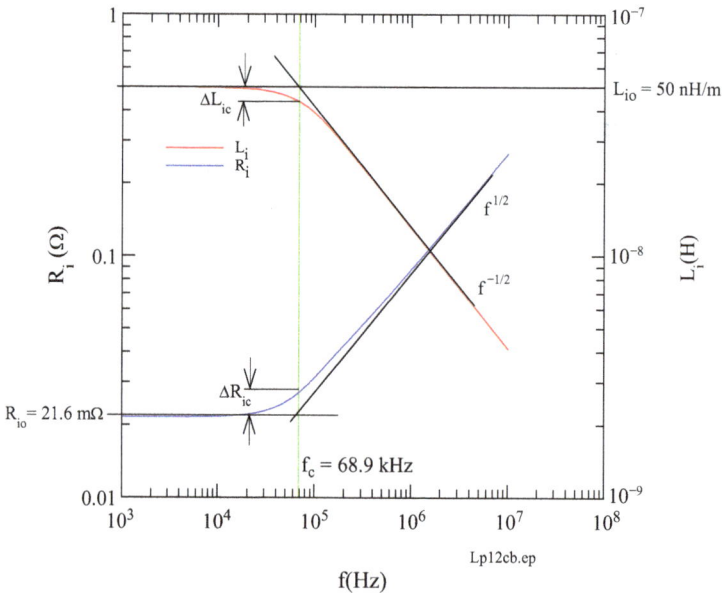

Lp12cb.ep

f(Hz)

Figure 7.4   Internal resistance R_i and inductance L_i frequency response for 1 mm diameter, 1 m long Cu wire. From the impedance Z_i (7.59) using Bessel functions. Sloping straight lines show the high-frequency approximations $R_i \propto f^{1/2}$ and $L_i \propto f^{-1/2}$. The horizontal lines indicate the constant zero frequency values $R_i = R_{io} = R_{dc}$, $L_i = L_{io} = 50$ nH/m. The constant frequency f_c occurs at the interception of the high-frequency approximations and the constant zero-frequency values.

Hence from (7.71)

$$f_c = \frac{R_{io}}{2\pi L_{io}} \tag{7.76}$$

For the general case of copper wires with different diameters (7.76), gives

$$f_c = \frac{R_{io}}{2\pi L_{io}} = \frac{\rho l}{\pi a^2 2\pi L_{io}} = \frac{0.01753\ l}{a^2} \tag{7.77}$$

For the case of Figure 7.4 with $a = 0.5$ mm, $l = 1$ m

$$f_c = \frac{\rho l}{\pi a^2 2\pi 50 e^{-9}} = 68.9\ \text{kHz}. \tag{7.78}$$

At this frequency R_i increases by 26% and L_i decreases by 12.9%.

In general the change in the inductance and resistance at f_c is given by

$$\Delta L_{ic} = L_{ic} - L_{io}, \quad \Delta R_{ic} = R_{ic} - R_{io} \tag{7.79}$$

where L_{ic} and R_{ic} are determined from (7.60) at the frequency f_c given by (7.76) or (7.77).

In (7.59), u_a is given by

$$u_a = j^{3/2}\sqrt{2}a/\delta \tag{7.80}$$

This may be re-written as

$$u_a = j^{3/2}\sqrt{f/f_{c1}} \tag{7.81}$$

where f is the normal applied frequency and

$$f_{c1} = \frac{\rho}{2\pi\mu a^2}, \quad Hz \tag{7.82}$$

At the frequency f_c described by (7.76) or (7.77), then

$$\frac{f_c}{f_{c1}} = 8 \tag{7.83}$$

and

$$u_{ac} = j^{3/2}\sqrt{8} = 2.828 j^{3/2} \tag{7.84}$$

Since generally u_a is given by (7.80), at the frequency f_c, $\sqrt{2}a/\delta = \sqrt{8}$. Thus, $a/\delta = \sqrt{2} = 1.414$.

For $a = 0.5$ mm, using a MATLAB program (Lipmat.txt), Table 7.1 lists the parameters which lead to the change in R_i and L_i at the frequency f_c.

7.7 Experimental measurements

The following results were obtained from copper, aluminium, iron, nickel and canthal wires measured using impedance analysers or a Gain Phase-Metre (GPM) method, [36] as stated in the figures.

The copper wire used was the type as used in standard house wiring, 99.90% pure copper [96,134]. The aluminium used was an 8 mm diameter aluminium alloy

Table 7.1   *The parameters which lead to the change in R_i and L_i at the frequency f_c*

a	0.5	mm
Rio	$2.1645e - 002$	Ω
Lio	$5.0000e - 008$	H
fc1	$8.6123e + 003$	Hz
fc	$6.8898e + 004$	Hz
Ric	$2.7373e - 002$	Ω
Lic	$4.3524e - 008$	H
dLi	$-6.4759e - 009$	H
dRi	$5.7282e - 003$	Ω
ΔL_i	$-1.2952e + 001$	%
ΔR_i	$2.6464e + 001$	%

rod with anodised surface, which was removed for the electrical contacts. This sample had a measured density of 2752 kgm^3 and resistivity of 30.78 $n\Omega m$. This compares with pure Al mean values of density 2703 kg m^3 and resistivity 26.7 $n\Omega m$, [97].

The iron wire was 99.5% pure. The nickel wire was Ni 95% and 5% (Al + Mn + Si). The Kanthal wire was the A1 alloy FeCrAl (22% Cr, 58% Al, 20%Fe). Resistivity $\rho = 1.45\mu\Omega.m$ at 20°C. Kanthal handbook p.15 [42,93].

7.8   Discussion and summary

Referring to the non-ferrous conductors, copper and aluminium, Figure 7.5 gives impedance measurements using the HP4192A impedance analyser on 8 mm dia aluminium alloy rod, total length 0.985 m, folded, gap 3 cm. The experimental results agree with theory for low frequencies, but the resistance begins to deviate at frequencies above 10 kHz, and the inductance at frequencies above about 200 kHz. In Figure 7.6 for the folded copper rod, a negative peak occurs in the internal inductance at about 100 kHz. It is not clear why this occurred and needs further investigation. However, the overall decrease in frequency agrees with the theoretical value of $f^{-1/2}$ Hz.

In the high frequency measurements, Figures 7.7 and 7.8, the R and L values look rather different, but the internal inductance dL_i is fairly close to the theoretical value of about 1 nH.

The measurements on the ferrous metals, Iron, Nickel and Kanthal wire, Figures 7.9–7.13, also showed clearly the transition to $f^{-1/2}$ for the internal inductance and $f^{1/2}$ for the resistance. Calculated values of the low-frequency permeability agreed approximately with expected values.

In the introduction, the reasons for this work were given mainly because of the lack of experimental data available on internal inductance measurements. But also

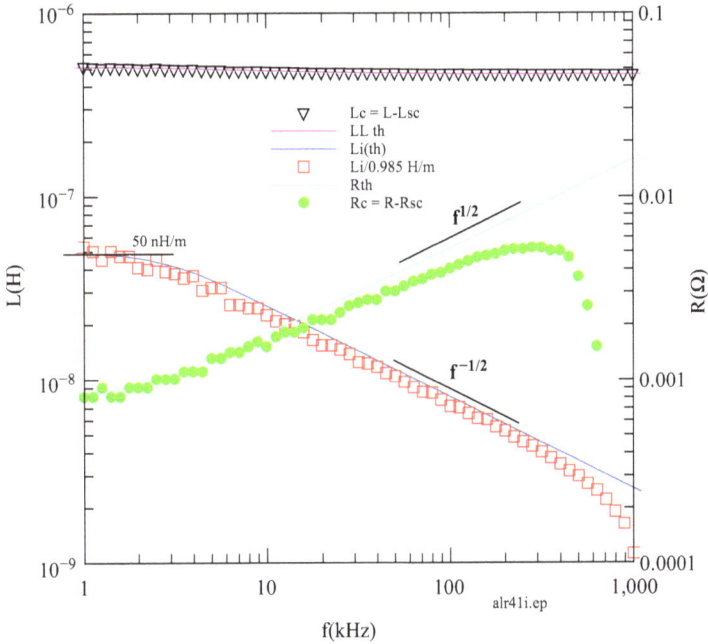

*Figure 7.5   Impedance measurements using HP4192A impedance analyzer on
8 mm dia aluminium alloy rod total length 0.985 m, folded, gap 3 cm.
Theoretical results indicated by (th).*

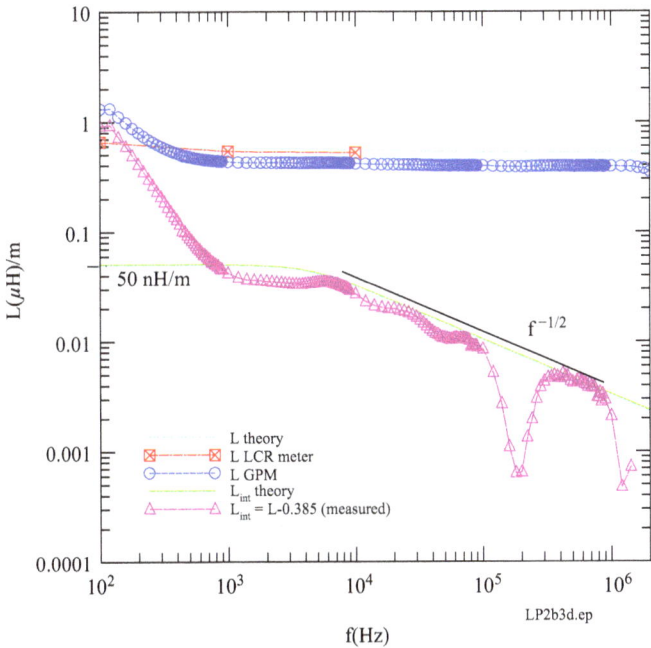

*Figure 7.6   GPM measurements on 4 mm dia Cu rod folded. Total length 0.85 m.
Zsc subtracted. Rs = 2.25 Ω,60 nH, Theory: MATLAB zrod.txt
Measurements: P2-2b.txt, P2-2sc3.txt, MATLAB Cumat.txt*

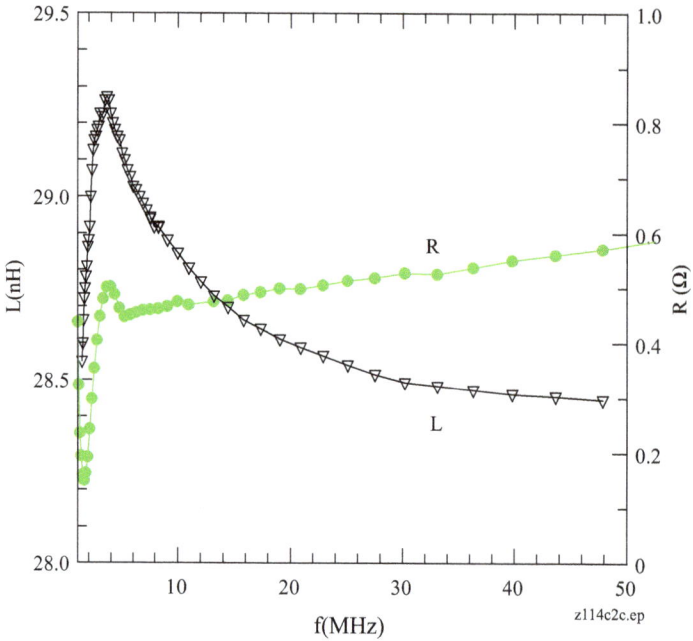

Figure 7.7 *R and L measurements of copper wire 0.114 mm diameter by 2 cm long using HP4191A impedance analyzer. Gives L = 28.404 nH at 47.95 MHz, L = 29.6 nH at 3.6 MHz, dL = Li = 1.2 nH. Theory gives Li = 1 nH.*

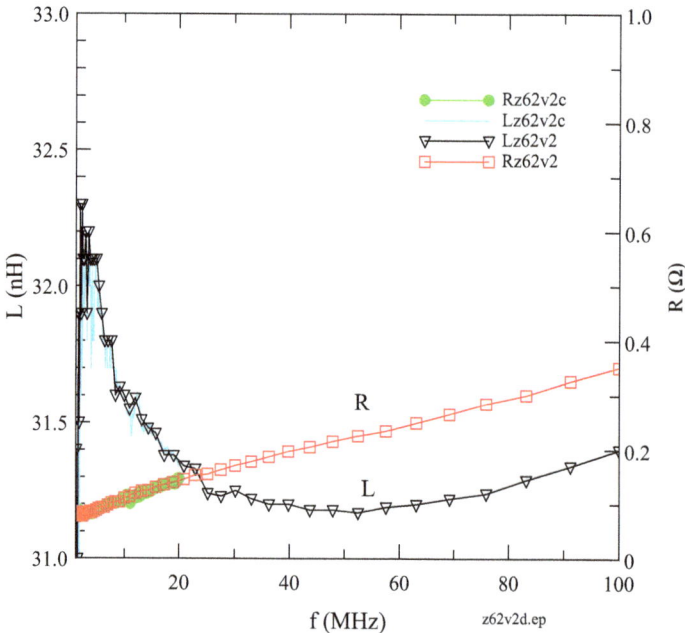

Figure 7.8 *R and L measurements of copper wire 0.062 mm diameter by 14.5 mm long using HP4191A impedance analyzer. dL = 32.1 nH (at 3.8 MHz) −31.17 nH (at 50 MHz) = 0.93 nH. Theory gives Li = 1 nH.*

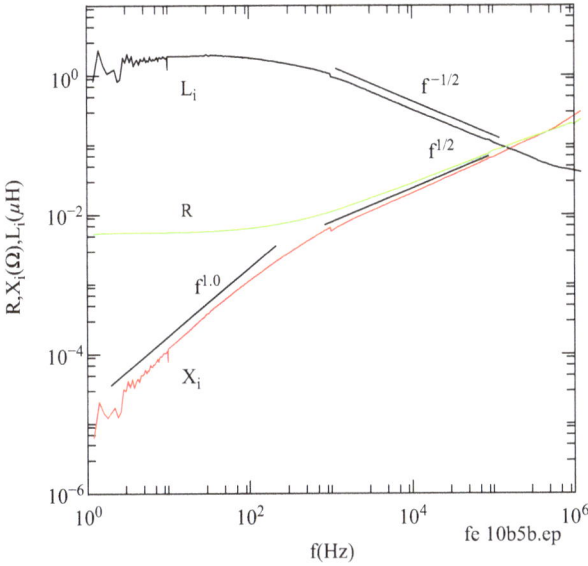

Figure 7.9 Iron wire, diameter 2 mm, formed into a rectangular loop with mean
gap width 18 mm, length 19.2 cm and current $I = 0.6$ Arms.
Measurements performed using a Gain Phase-Meter (GPM)
method [36] with function generator HP3325A and low-frequency
power amplifier for frequency range 10 Hz to 1 kHz and
high-frequency power amplifier for 1 kHz to 1 MHz. Internal
inductance (L_i), resistance (R) and reactance (X_i).

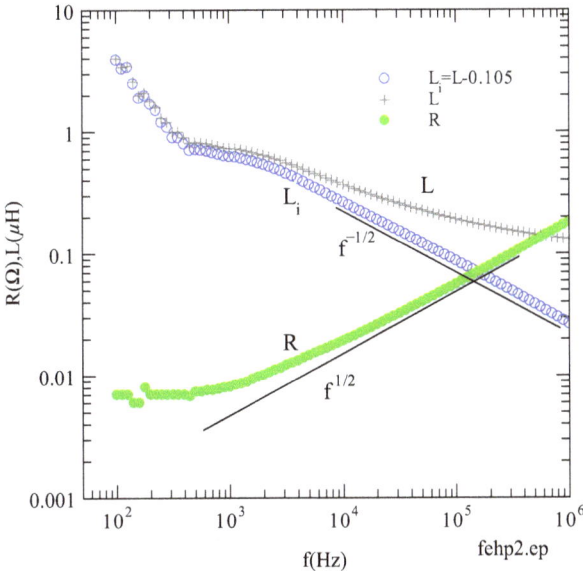

Figure 7.10 Iron wire as shown in Figure 7.9. Measurements performed using
Impedance Analyzer (HP4192A). Inductance (L), internal inductance
(L_i) and resistance (R).

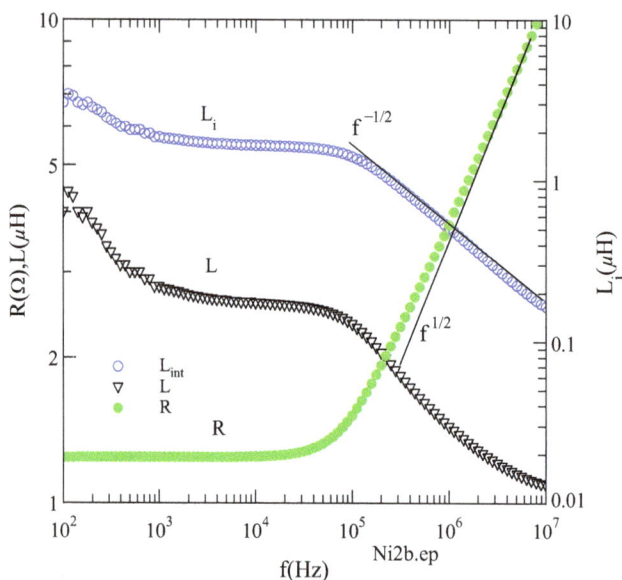

Figure 7.11 Nickel wire 0.5 mm diameter and 45.5 cm long. Measurements performed using Impedance Analyzer (HP4192A). Inductance (L), internal inductance ($L_i = L - 0.923~\mu H$) and resistance (R).

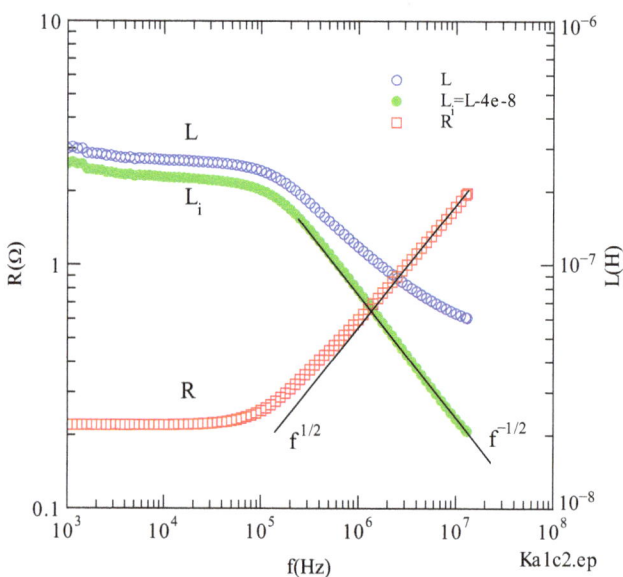

Figure 7.12 Kanthal wire 0.68 mm diameter, 5 cm long. Measurements performed using Impedance Analyzer (HP4192A). Inductance (L), internal inductance (L_i) and resistance (R).

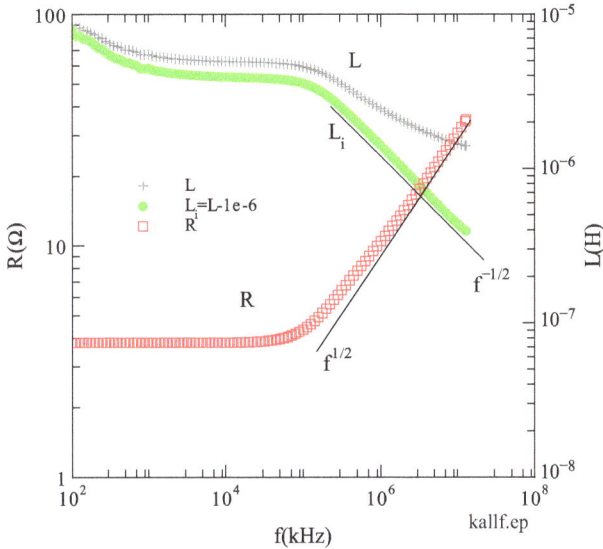

Figure 7.13   Kanthal wire 0.68 mm diameter, total electrical length 98 cm, formed
in to a rectangular loop gap 5 cm. Measurements performed using
Impedance Analyzer (HP4192A). Inductance (L), internal inductance
(L_i) and resistance (R).

because these measurements can supply an alternative measurement of the permeability of conductors, particularly in ferrous metals. This is followed by the analysis of the internal inductance of cylindrical conductors for both steady state d.c. and sinusoidal a.c. sources. The ohmic loss and internal ac resistance are then determined from power dissipation in the conductor. In each case of the internal inductance and resistance, the low frequency limit is obtained and compared with the steady state calculations. Experimental measurements of the internal inductance and resistance of non-ferrous and ferrous metals are then presented. Although the measurements on non-ferrous metals, copper and aluminium, were weak at low frequencies due to the unity relative permeability, the results obtained revealed the low frequency threshold of 50 nH/m and the high frequency decrease proportional to $f^{-1/2}$ Hz. Generally, the impedance measurement techniques employed have successfully revealed the internal inductance of non-ferrous and ferrous metals, but agreement with theory is not always held.

7.9   Appendix

7.9.1   *Internal inductance from the internal magnetic flux*

The internal inductance may also be determined from the ratio of the internal magnetic flux Ψ to the ac current $I(\omega)$ according to

$$L_i = \frac{\Psi}{I(\omega)} = \frac{1}{I(\omega)} \int_s B_\phi ds = \frac{l}{I(\omega)} \int_0^a B_\phi dr \qquad (7.85)$$

where the integration is along the length l of the conductor over an area between 0 and a. Substituting for the current (7.42) and (B_ϕ) from (5.125) gives

$$L_i = \frac{\mu l}{2\pi a B_\phi} \int_0^a B_\phi dr = \frac{\mu l}{2\pi a J_1(u_{1a})mj^{3/2}} \int_0^{u_{1a}} J_1(u_1)du \tag{7.86}$$

Using the Bessel identity

$$\int_0^{u_a} J_1(u)du = 1 - J_o(u_a) \tag{7.87}$$

then

$$L_i = \frac{\mu l}{2\pi a} \frac{1 - J_o(u_{1a})}{u_{1a}J_1(u_{1a})} \tag{7.88}$$

At low frequencies

$$J_o(u_{1a}) = 1 - u_{1a}^2, J_1(u_{1a}) = u_{1a}/2 \tag{7.89}$$

leading to

$$L_i = \frac{\mu l}{4\pi} \tag{7.90}$$

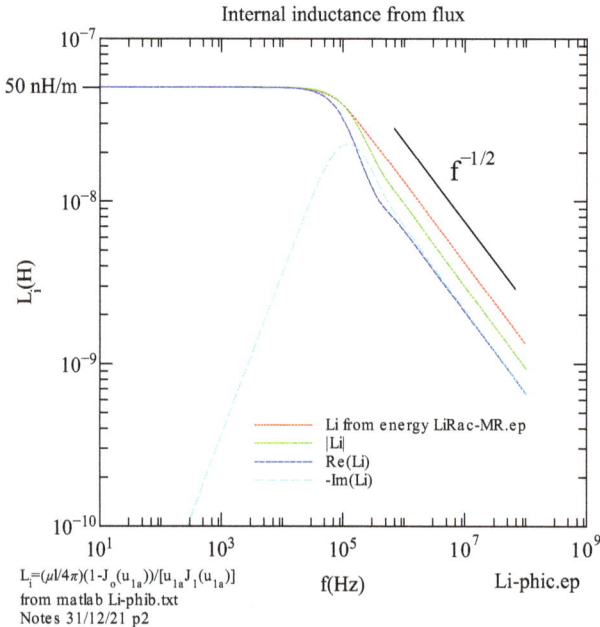

Internal inductance from flux

L$_i$(H)

Li from energy LiRac-MR.ep
|Li|
Re(Li)
-Im(Li)

$L_i=(\mu l/4\pi)(1-J_o(u_{1a}))/[u_{1a}J_1(u_{1a})]$
from matlab Li-phib.txt
Notes 31/12/21 p2

f(Hz)   Li-phic.ep

$f^{-1/2}$

Figure 7.14   Internal inductance from magnetic flux for copper wire, radius 0.5 mm, length 1.0 m, I = 1.0 A

This is the result for an infinite wire or twin transmission lines. For a sem-infinite wire or single wire $I = 4\pi aB/\mu$. Then

$$L_i = \frac{\mu l}{4\pi a} \frac{1 - J_o(u_{1a})}{u_{1a}J_1(u_{1a})} \tag{7.91}$$

At low frequencies this gives $L_i = \frac{\mu l}{8\pi}$ as derived previously (7.66). Figure 7.14 shows a plot of the inductance L_i from (7.91) using MATLAB.

Chapter 8
Cylindrical conductors – axial AC magnetic field

8.1 Introduction

The application of a time-varying magnetic field is widely used in electromagnetics in devices such as transformers, electric motors, generators, eddy current testing and other non-contact measuring techniques. In these cases, the induced voltage and currents may be determined using Faraday's law, induced $emf = -d\phi/dt$, where ϕ is the flux threading the conductors. However, this approach does not reveal how the magnetic field penetrates the conductors and its spatial variation within the conductor. A more detailed approach is to solve the time-dependent Maxwell's equations.

8.2 Magnetic field penetration

In the present case, we consider a time-dependent magnetic field $\mathbf{B} = \mathbf{B}_o e^{j\omega t}$ applied axially to a long cylindrical conductor with radius a, Figure 8.1. $\mathbf{B}_o$ is less than the saturation flux density $\mathbf{B}_{sat}$ and any static magnetic field is assumed zero, that is, $\mathbf{B}_{dc} = 0$.

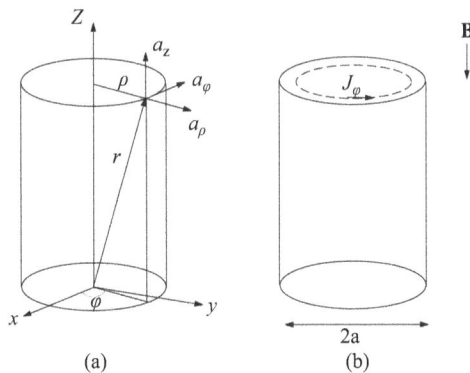

Figure 8.1   *(a) General cylindrical co-ordinates and (b) relationship to a cylindrical conductor in a magnetic field B (cyl4.eps)*

The field penetration into the conductor is obtained from Maxwell's equations, which yield the diffusion (5.3), re-written here as

$$\nabla^2 \mathbf{B} = jm^2 \mathbf{B} \tag{8.1}$$

where $m^2 = \omega\mu\sigma$ or $m = \sqrt{2}/\delta$, where the skin depth $\delta = \sqrt{(2/\omega\mu\sigma)}$. Within the conductor, we seek solutions of the form

$$\mathbf{B}(r,t) = \mathbf{B}(r)e^{j\omega t} \tag{8.2}$$

At the cylinder surface $r = a$, the field is just the applied field $\mathbf{B}(a,t) = \mathbf{B}_o e^{j\omega t}$. Equation (8.1) may be expressed in cylindrical co-ordinates, Figures 8.1(a) and 8.1(b), where in this case we can take $\rho = r$.

$$\nabla^2 B = r^{-1}\frac{\partial}{\partial r}\left(r\frac{\partial B}{\partial r}\right) + r^{-2}\frac{\partial^2 B}{\partial \phi^2} + \frac{\partial^2 B}{\partial z^2} = jm^2 B. \tag{8.3}$$

Assuming B does not vary with angle ϕ or distance z, then

$$r^2\frac{\partial^2 B}{\partial r^2} + r\frac{\partial B}{\partial r} - jm^2 B r^2 = 0 \tag{8.4}$$

Bessel's modified equation order v is defined by

$$u^2\frac{\partial^2 B}{\partial u^2} + u\frac{\partial B}{\partial u} + (u^2 - v^2)B = 0 \tag{8.5}$$

where $u = mrj\sqrt{j}$ and in our case $v = 0$. The general solution is

$$B(r) = AJ_o(u) \tag{8.6}$$

where $J_o(u)$ is the Bessel function of the first kind with complex argument u and order zero.

See the general Appendix, Bessel's modified equation.

At the boundary of the conductor, $r = a$, $B(a) = AJ_o(u_a)$, where $u_a = amj\sqrt{j}$. Taking the ratio of these two fields eliminates A. Hence, substituting $B(r)$ in (8.2) gives for the magnetic flux density within the conductor

$$B(r,t) = B(a)\frac{J_o(u)}{J_o(u_a)}e^{j\omega t} \tag{8.7}$$

Figure 8.2 shows plots of the modulus, real and imaginary normalised amplitude terms $B_N = B(r)/B(a)$ for (8.7), using the Bessel functions available in MATLAB®. Figure 8.2(a) shows that the modulus $|B_N|$ becomes flat and nearly equal to the applied field $B(a)$ as the frequency decreases. As the frequency increases, Figure 8.2(b), the flux density decreases towards the centre of the conductor.

8.2.1   Flux density complex components

Equation (8.7) can be re-expressed in terms of real and imaginary complex components, sometimes referred to as Kelvin functions [78].

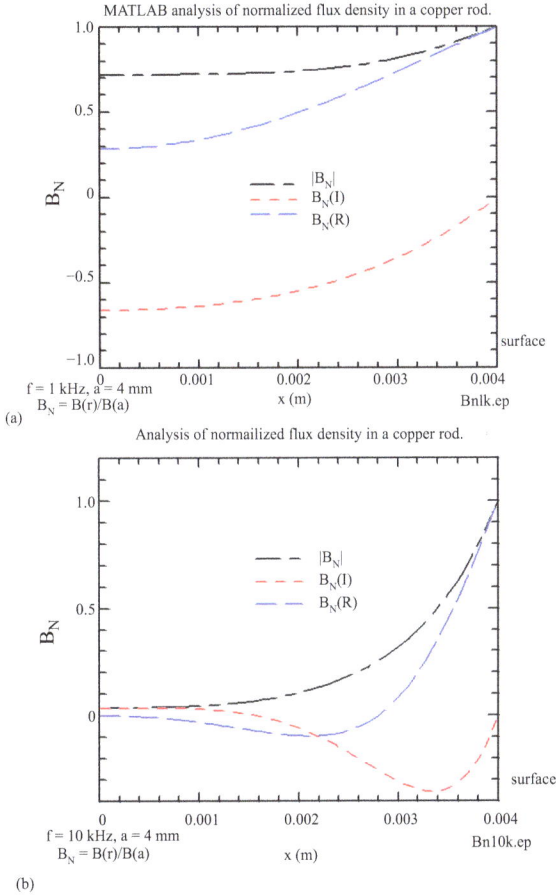

Figure 8.2 *Plot of (8.7) normalised amplitude. Copper rod, radius 4 mm, (a) f = 1 kHz and (b) f = 10 kHz*

Substituting for $u = rmj\sqrt{j}$ in (8.5), where $m = \sqrt{\omega\mu\sigma} = \sqrt{2}/\delta$, gives (8.4). Hence,

$$J_o(u) = 1 + j(mr/2)^2 - \frac{(mr/2)^4}{(2!)^2} - j\frac{(mr/2)^6}{(3!)^2} + \cdots \tag{8.8}$$

Separating out the real and imaginary complex components gives the Kelvin functions

$$RJ_o(u) = ber_o(mr) = 1 - \frac{(mr/2)^4}{(2!)^2} + \frac{(mr/2)^8}{(4!)^2} - \frac{(mr/2)^{12}}{(6!)^2} + \cdots \tag{8.9}$$

$$IJ_o(u) = bei_o(mr) = (mr/2)^2 - \frac{(mr/2)^6}{(3!)^2} + \frac{(mr/2)^{10}}{(5!)^2} \cdots \tag{8.10}$$

Hence,

$$B(mr) = A[ber_o(mr) + jbei_o(mr)] \tag{8.11}$$

At the boundary $r = a$, where $B(mr) = B(ma) = const$, then

$$B(ma) = A[ber_o(ma) + jbei_o(ma)] \tag{8.12}$$

Substituting for A

$$B(mr) = B(ma)\frac{ber_o(mr) + jbei_o(mr)}{ber_o(ma) + jbei_o(ma)} \tag{8.13}$$

Rationalising and substituting into (8.2) gives for the magnetic field within the conductor

$$B(mr,t) = B(ma)\sqrt{\frac{ber_o^2(mr) + bei_o^2(mr)}{ber_o^2(ma) + bei_o^2(ma)}}e^{j(\omega t + \phi_1)} \tag{8.14}$$

where the phase difference ϕ_1 is

$$\phi_1 = tan^{-1}\frac{bei_o(mr)}{ber_o(mr)} - tan^{-1}\frac{bei_o(ma)}{ber_o(ma)} \tag{8.15}$$

In this approach, the amplitude and phase are given explicitly by (8.14) and (8.15), whereas in (8.7) the amplitude and phase are embedded in the Bessel functions.

8.3   The average permeability

The average permeability is defined by

$$\mu = \frac{\bar{B}}{B_a} = \frac{2}{a^2 B_a}\int_0^a rB(r)dr \tag{8.16}$$

where at the cylinder surface $B_a = B(a) = \mu_o H_a$. Thus, if the magnetic field in the specimen B_{int} is uniform and constant, then the average internal field is $\bar{B} = (2/a^2)B_{int}[a^2/2] = B_{int}$ as expected. Substituting for the amplitude $B(r)$ in (8.7) gives

$$\mu = \frac{2}{a^2}\int_0^a r\frac{J_o(u)}{J_o(u_a)}dr \tag{8.17}$$

where

$$u = mrj^{3/2}, \quad r = uj^{-3/2}/m, \quad dr/du = j^{-3/2}/m, \quad u_a = maj^{3/2}. \tag{8.18}$$

Hence,

$$\mu = \frac{2}{u_a^2}\int_0^{u_a} u\frac{J_o(u)}{J_o(u_a)}du \tag{8.19}$$

Using the identity

$$\int uJ_o(u)du = uJ_1(u) \tag{8.20}$$

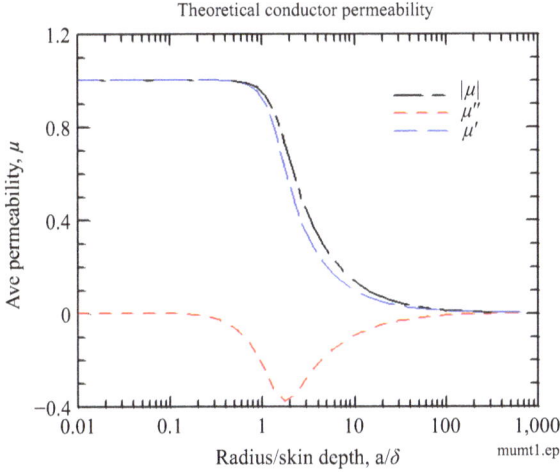

Figure 8.3   Average permeability from (8.21) as a function of the ratio conductor radius to skin depth, a/δ. (mumt1.eps)

gives the average permeability of the conductor as

$$\mu = \frac{2}{u_a} \frac{J_1(u_a)}{J_0(u_a)} \tag{8.21}$$

Figure 8.3 shows a plot of (8.21) as a function of the ratio conductor radius to skin depth, a/δ. This latter parameter is independent of frequency and conductivity. The curves are therefore universal and should apply to any conductor within the limitations of the model used.

8.3.1   Complex permeability

$$\mu = \frac{2}{a^2} \int_0^a r \frac{ber_o(mr) + jbei_o(mr)}{ber_o(ma) + jbei_o(ma)} \, dr = \frac{\overline{ber_o}(mr) + j\overline{bei_o}(mr)}{ber_o(ma) + jbei_o(ma)} \tag{8.22}$$

$$\overline{ber_o} = \frac{2}{a^2} \int_0^a rber_o(mr)dr = 1 - \frac{\alpha^4}{3(2!)^2} + \frac{\alpha^8}{5(4!)^2} - \frac{\alpha^{12}}{7(6!)^2} + \cdots \tag{8.23}$$

$$\overline{bei_o} = \frac{2}{a^2} \int_0^a rbei_o(mr)dr = \frac{\alpha^2}{2} - \frac{\alpha^6}{4(3!)^2} + \frac{\alpha^{10}}{6(5!)^2} + \cdots \tag{8.24}$$

where $\alpha = ma/2$. Hence,

$$\mu = \frac{(ber_o\overline{ber_o} + bei_o\overline{bei_o}) - j(bei_o\overline{ber_o} - ber_o\overline{bei_o})}{ber_o^2 + bei_o^2} = \mu' - j\mu'' \tag{8.25}$$

Note that μ'' is due to energy loss and is negative because the device is passive, that is, there is no energy gain.

8.3.2   Normal and superconducting cylinders

Both Landau *et al.* [12] and Gormory [80] give the same solution for the suscepti-bility of cylindrical normal conductors and superconductors in parallel ac fields. In this work, the authors tend to use $kr = mrj\sqrt{j} = -(1-j)r/\delta$ for the Bessel function arguments. Hence the amplitude in (8.7) becomes

$$B(r) = B_a \frac{J_o(kr)}{J_o(ka)} \tag{8.26}$$

The permeability is then

$$\mu = \frac{\overline{B(r)}}{B_a} = \frac{2}{a^2 k^2 J_o(ka)} \int_0^a kr J_o(kr) \, d(kr) \tag{8.27}$$

Using $\int u J_o(u) du = u J_1(u)$ gives

$$\mu = \frac{2}{ak} \frac{J_1(ka)}{J_o(ka)}, \quad \mu_a' = Re\left[\frac{2}{u_a} \frac{J_1(u_a)}{J_o(u_a)}\right], \quad \mu_a'' = Im\left[\frac{2}{u_a} \frac{J_1(u_a)}{J_o(u_a)}\right] \tag{8.28}$$

Substituting for the Bessel series gives

$$\mu_a = \frac{2}{u_a} \frac{J_1(u_a)}{J_o(u_a)} = \frac{2}{u_a} \frac{\frac{u_a}{2} - \frac{(u_a/2)^3}{1!2!} + \frac{(u_a/2)^5}{2!3!} - \frac{(u_a/2)^7}{3!4!} + \cdots}{1 - \frac{(u_a/2)^2}{(1!)^2} + \frac{(u_a/2)^4}{(2!)^2} - \frac{(u_a/2)^6}{(3!)^2} + \cdots} \tag{8.29}$$

where $u/2 = (ma/2)j\sqrt{j} = \alpha j\sqrt{j}$ and $(j\sqrt{j})^2 = -j$, $(j\sqrt{j})^4 = -1$, $(j\sqrt{j})^6 = j$, $(j\sqrt{j})^8 = 1$, $(j\sqrt{j})^{10} = -j$, $(j\sqrt{j})^{12} = -1\dots$ gives

$$\mu = \frac{1 + j\frac{\alpha^2}{2!} - \frac{\alpha^4}{2!3!} - j\frac{\alpha^6}{3!4!} + \frac{\alpha^8}{4!5!} + j\frac{\alpha^{10}}{5!6!} - \frac{\alpha^{12}}{6!7!} + \cdots}{1 + j\frac{\alpha^2}{1!} - \frac{\alpha^4}{(2!)^2} - j\frac{\alpha^6}{(3!)^2} + \frac{\alpha^8}{(4!)^2} + j\frac{\alpha^{10}}{(5!)^2} - \frac{\alpha^{12}}{(6!)^2} + \cdots}, \quad \alpha = ma/2 \tag{8.30}$$

$$\mu = \frac{(1 - \frac{\alpha^4}{2!3!} + \frac{\alpha^8}{4!5!} - \cdots) + j(\frac{\alpha^2}{2!} - \frac{\alpha^6}{3!4!} + \frac{\alpha^{10}}{5!6!} - \cdots)}{(1 - \frac{\alpha^4}{(2!)^2} + \frac{\alpha^8}{(4!)^2} - \cdots) + j(\frac{\alpha^2}{1!} - \frac{\alpha^6}{(3!)^2} + \frac{\alpha^{10}}{(5!)^2} - \cdots)} \tag{8.31}$$

These equations are identical to those obtained previously, (8.22) to (8.24). The real and imaginary parts are then obtained from (8.25). Note that $2!3! = 3(2!)^2, 4!5! = 5(4!)^2$, etc.

8.3.3   Approximations

Using the first and second terms only of the ber_o and bei_o series gives

$$\mu = \frac{(1 - \frac{\alpha^4}{2!3!}) + j(\frac{\alpha^2}{2!})}{(1 - \frac{\alpha^4}{(2!)^2}) + j(\frac{\alpha^2}{1!})} = \frac{(1 - \frac{\alpha^4}{12}) + j\frac{\alpha^2}{2}}{(1 - \frac{\alpha^4}{4}) + j\frac{\alpha^2}{1}} \tag{8.32}$$

$$\mu = \frac{(1 + \frac{\alpha^4}{6} + \frac{\alpha^8}{48}) - j\frac{\alpha^2}{2}(1 + \frac{\alpha^4}{12})}{1 + \frac{\alpha^4}{2} + \frac{\alpha^8}{16}} \tag{8.33}$$

Hence,

$$\chi' = \mu' - 1 = -\frac{\left(\frac{\alpha^4}{3} + \frac{\alpha^8}{24}\right)}{1 + \frac{\alpha^4}{2} + \frac{\alpha^8}{16}} \approx -\frac{1}{12}(a/\delta)^4 \tag{8.34}$$

$$\mu'' = -\chi'' = \frac{\frac{\alpha^2}{2}\left(1 + \frac{\alpha^4}{12}\right)}{1 + \frac{\alpha^4}{2} + \frac{\alpha^8}{16}} \approx \alpha^2/2 = \frac{1}{4}(a/\delta)^2 \tag{8.35}$$

Using $\delta = \sqrt{\frac{2\rho}{\omega\mu}}$, then

$$\rho' = \frac{\omega\mu_o a^2}{2\sqrt{12\chi'}}, \quad \rho'' = \frac{\omega\mu_o a^2}{8\chi'}, \tag{8.36}$$

These results are equivalent to those obtained by Landau for a long conducting cylinder in a magnetic field parallel to the cylinder axis and $\delta \gg a$. (converted to SI units by multiplying Landau's results by 4π). Putting $\rho_1 = \omega\mu_o a^2/2$, then

$$\rho' = \frac{\rho_1}{\sqrt{12\chi'}}, \quad \rho'' = \frac{\rho_1}{4\chi''}, \tag{8.37}$$

If the theoretical and experimental resistivities are ρ_{th} and ρ_{exp}, respectively, then

$$\rho'_{th} = \frac{\rho_1}{\sqrt{12\chi'_{th}}}, \quad \rho''_{th} = \frac{\rho_1}{4\chi''_{th}}, \tag{8.38}$$

$$\rho'_{exp} = \frac{\rho_1}{\sqrt{12\chi'_{exp}}}, \quad \rho''_{exp} = \frac{\rho_1}{4\chi''_{exp}}, \tag{8.39}$$

$$\frac{\rho'_{exp}}{\rho'_{th}} = \sqrt{\frac{\chi'_{th}}{\chi'_{exp}}}, \quad \frac{\rho''_{exp}}{\rho''_{th}} = \frac{\chi''_{th}}{\chi''_{exp}} \tag{8.40}$$

Hence, error in $\rho'_{exp} \propto (1/\sqrt{\chi'_{exp}}$ and error in $\rho''_{exp} \propto 1/\chi''_{exp}$.

8.3.4 Current density and electric field

The current density may be determined from Ampere's law or the point form of Maxwell's equation $curl\mathbf{H} = \partial\mathbf{D}/\partial t + \mathbf{J}$. Neglecting the displacement current and using cylindrical co-ordinates noting that H_z is only in the negative z-direction

$$\mathbf{H} = 0\mathbf{a}_\rho + 0\mathbf{a}_\phi - H_z\mathbf{a}_z \tag{8.41}$$

$$\nabla \times \mathbf{H} = \left(\frac{1}{\rho}\frac{\partial(-H_z)}{\partial\phi} - \frac{\partial H_\phi}{\partial z}\right)\mathbf{a}_\rho + \left(\frac{\partial H_\rho}{\partial z} - \frac{\partial(-H_z)}{\partial\rho}\right)\mathbf{a}_\phi$$

$$+ \frac{1}{\rho}\left(\frac{\rho\partial H_\phi}{\partial\rho} - \frac{\partial H_\rho}{\partial\phi}\right)\mathbf{a}_z \tag{8.42}$$

The terms containing H_ρ and H_ϕ are zero. Hence,

$$\nabla \times \mathbf{H} = \mathbf{J} = -\frac{1}{\rho}\frac{\partial H_z}{\partial\phi}\mathbf{a}_\rho + \frac{\partial H_z}{\partial\rho}\mathbf{a}_\phi \tag{8.43}$$

H_z does not vary with ϕ. Put $\rho = r$ and $H_z = B_z/\mu$ where $\mu = \mu_0\mu_r$. Then, (8.43) becomes

$$J_\phi = \frac{1}{\mu}\frac{\partial B(r)}{\partial r} \tag{8.44}$$

Differentiating the amplitude in (8.7), then

$$\frac{dB(r)}{dr} = \frac{B_a}{J_o(u_a)}\frac{dJ_o(u)}{dr} \tag{8.45}$$

Using the identity

$$\frac{dJ_o(u)}{du} = -J_1(u) \tag{8.46}$$

and the r to u substitutions in (8.18) and noting that $-j^{3/2} = j^{1/2}$, we obtain

$$J_\phi(r) = \frac{j^{1/2}mB_a}{\mu}\frac{J_1(u)}{J_o(u_a)} \tag{8.47}$$

Assuming $J_\phi = \sigma E_\phi$, then the electric field, $E_\phi = \rho J_\phi$, is

$$E_\phi(r) = \frac{j^{1/2}mB_a\rho}{\mu}\frac{J_1(u)}{J_o(u_a)} \tag{8.48}$$

The theoretical current density across the radius of a copper rod is shown in Figure 8.4. (a) Shows a plot of (8.47) for the current density modulus $|J|$ for $B_a = 1\,\text{mT}$, copper rod radius $4\,\text{mm}$ at several different frequencies. The current density is seen to fall from about $10^5\,\text{Am}^{-2}$ at the surface to less than $10^{-15}\,\text{Am}^{-2}$ at the conductor centre for the 1 MHz case, a decrease of more than 20 decades. This result is mirrored on the negative x-axis in this two-dimensional sectional plot.

The electric field determined for $r = 0$ to $r = a$, (see Appendix A) is

$$E_{\phi a} = \frac{j^{1/2}\omega B_a}{m}\frac{J_1(u_a)}{J_o(u_a)} \tag{8.49}$$

Hence, the current density is

$$J_{\phi a} = \frac{j^{1/2}mB_a}{\mu}\frac{J_1(u_a)}{J_o(u_a)} \tag{8.50}$$

8.3.5   Current density Kelvin functions

In terms of the Kelvin functions the current density equation becomes

$$J_\phi(r,t) = -\frac{\sqrt{2}B_a}{\delta\mu}\sqrt{\frac{ber_o'^2(mr) + bei_o'^2(mr)}{ber_o^2(ma) + bei_o^2(ma)}}e^{j(\omega t+\phi_2)} \tag{8.51}$$

The phase angle between current and magnetic field is

$$\phi_2 = \frac{3\pi}{4} + tan^{-1}\frac{bei_o'(mr)}{ber_o'(mr)} - tan^{-1}\frac{bei_o(ma)}{ber_o(ma)} \tag{8.52}$$

The prime (ı) in *ber'* and *bei'* refers to the Bessel function order 1.

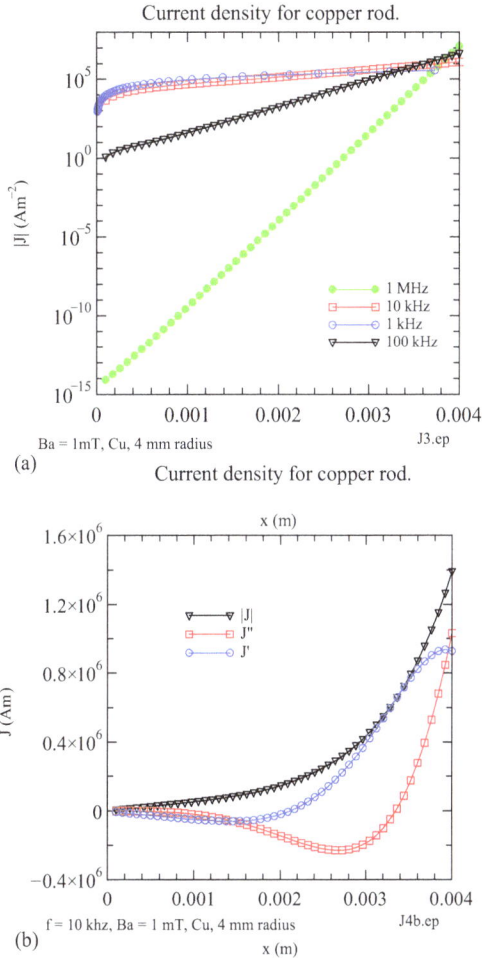

Figure 8.4 (a) Current density across the radius of a copper rod for several frequencies according to (8.47). Applied field amplitude $B_a = 1\,mT$, copper rod radius 4 mm. (b) As (a) but complex components of J at 10 KHz.

8.4 Total current

The total current is obtained by considering a small area element bdr, where b is the cylinder length, Figure 8.5. The total current amplitude is then

$$I = \int_s \mathbf{J}.ds = b \int_0^a J_\phi dr \tag{8.53}$$

Substituting for (8.47) gives

$$I = -\frac{mj\sqrt{j}B_a b}{\mu_o} \int_0^a \frac{J_1(mrj\sqrt{j})}{J_o(maj\sqrt{j})} dr \tag{8.54}$$

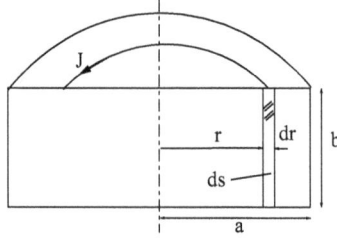

Figure 8.5   Calculation of the total azimuthal current from a section of the conductor

where $u = mrj^{3/2}$, $r = uj^{-3/2}/m$, $dr/du = j^{-3/2}/m$, $u_a = maj^{3/2}$, then

$$I = -\frac{B_a b}{\mu_o J_o(u_a)} \int_0^{u_a} J_1(u)du \tag{8.55}$$

Substituting for the identity

$$\int_0^{u_a} J_1(u)du = 1 - J_o(u_a) \tag{8.56}$$

and inserting the time element gives

$$I(t) = -\frac{B_a b}{\mu_o} \frac{[1 - J_o(u_a)]}{J_o(u_a)} e^{j\omega t} \tag{8.57}$$

The Bessel functions are dimensionless. The coefficient $B_a b/\mu$ has dimensions $Am^{-1}m = A$ as expected. It is also useful to obtain the current profile across the conductor. In this case using the indefinite integral, we obtain for the current amplitude

$$I(r) = -\frac{B_a b}{\mu_o} \frac{[1 - J_o(u)]}{J_o(u_a)} \tag{8.58}$$

8.5   Induced voltage

The emf induced into a conducting loop, Figures 8.6(a) and 8.6(b), due to the perpendicular varying magnetic field is

$$V(r,t) = -\frac{d\Phi}{dt} = -\frac{d}{dt} \int_s B(r)e^{j\omega t}.ds \tag{8.59}$$

Hence, putting $ds = rdrd\phi$, Figure 8.6(c), the amplitude is

$$V(r) = -j\omega \int_o^a \int_0^{2\pi} rB(r)d\phi dr. \tag{8.60}$$

$B(r)$ is in the negative z direction. It is uniform in ϕ varying spatially only with r. Substituting for $B(r)$ (8.7) into (8.60) gives

$$V(r) = j\omega 2\pi B_a \int_o^a r\frac{J_o(u)}{J_o(u_a)}dr \tag{8.61}$$

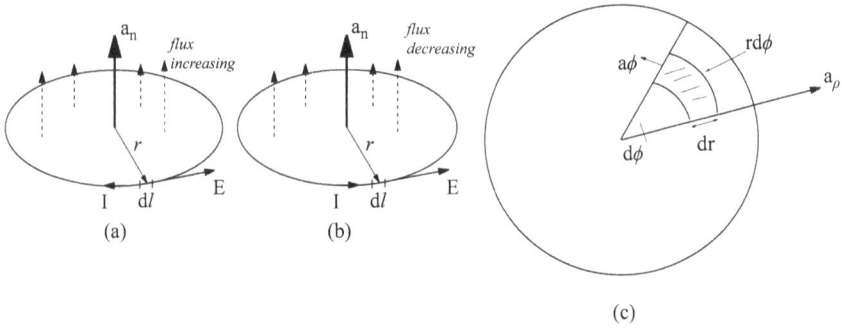

Figure 8.6 *Induced current and electric field in a conducting loop due to a varying magnetic field normal to the loop. (a) Flux increasing, (b) flux decreasing, (c) loop geometry*

For $u = mrj^{3/2}, r = uj^{-3/2}/m, dr/du = j^{-3/2}/m, u_a = maj^{3/2}$, then

$$V(r) = j\omega 2\pi B_o a^2 \int_0^a \frac{u J_o(u)}{u_a^2 J_o(u_a)} du \qquad (8.62)$$

Using the identity: $\int u J_o(u) du = u J_1(u)$, then

$$V(r) = j\omega 2\pi B_a a^2 \left[\frac{u J_1(u)}{u_a^2 J_o(u_a)} \right]_0^{u_a} \qquad (8.63)$$

Recalling the time element this becomes

$$V(a, t) = j\omega \pi a^2 B_a \frac{2 J_1(u_a)}{u_a J_o(u_a)} e^{j\omega t} \qquad (8.64)$$

The bracketed term is found to be the average permeability (8.21). Hence,

$$V(a, t) = j\pi a^2 \omega B_a < \mu > e^{j\omega t} \qquad (8.65)$$

To see how $V(r)$ varies across the conductor, we use the indefinite integral in (8.62) to give

$$V(r) = j\omega 2\pi B_a a^2 \frac{u J_1(u)}{u_a^2 J_o(u_a)}. \qquad (8.66)$$

Substitute for $u_a^2 = (maj^{3/2})^2$ and $m^2 = \omega\mu_o\sigma$, then

$$V(r) = -\frac{2\pi\rho B_a}{\mu_o} \frac{u J_1(u)}{J_o(u_a)} \qquad (8.67)$$

8.5.1 Induced voltage from the electric field

The current and electric field induced in the loop, Figures 8.6 and 8.7, due to the perpendicular varying magnetic field is

$$V(r) = \oint_l \mathbf{E}.d\mathbf{l} \qquad (8.68)$$

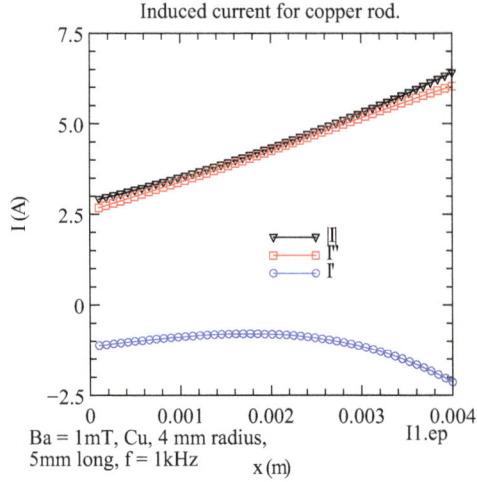

Figure 8.7   Total current profile for a copper rod using (8.58). Applied field amplitude $B_a = 1$ mT, copper rod radius 4 mm, length 5 mm, 1 kHz

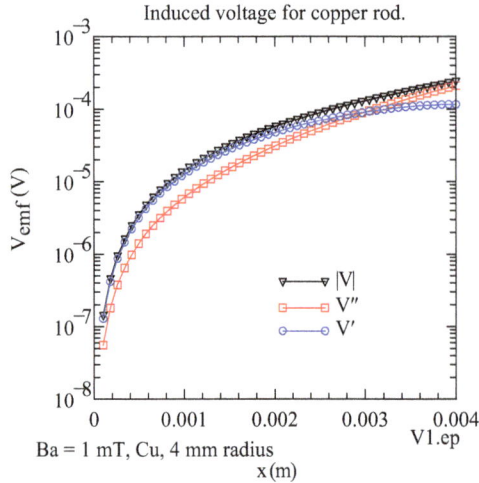

Figure 8.8   Induced voltage in a cylindrical conductor due to a varying axial magnetic field. Applied field amplitude $B_a = 1$ mT, $f = 1$ kHz, copper rod radius 4 mm

where the integral is around the closed path of the loop. Substituting for the azimuthal electric field (8.48) and integrate around the closed loop gives

$$V(r) = \int_0^{2\pi} \frac{j\omega j^{-3/2} B_a}{m} \frac{J_1(u)}{J_o(u_a)} r d\phi \tag{8.69}$$

$$V(r) = \frac{j^{-1/2} 2\pi \omega B_a}{m} \frac{r J_1(u)}{J_o(u_a)} \tag{8.70}$$

As before use substitutes $u = mrj^{3/2}$ and $m^2 = \omega\mu_o\sigma$, then

$$V(r) = -\frac{2\pi\rho B_a}{\mu_o} \frac{u J_1(u)}{J_o(u_a)} \tag{8.71}$$

This agrees with (8.67). A graph of (8.71) is shown in Figure 8.8 for the real, imaginary and modulus of V varying across the radius of the conductor. At this frequency, 1 kHz, these voltages decrease monotonically from the surface to the centre of the conductor.

Chapter 9

Cylindrical conductors in axial magnetic fields – impedance

9.1 Impedance from *V/I*

From the previous calculations of voltage and current we can also obtain the impedance $Z = V/I$ from the ratio of (8.71) and (8.58). This gives

$$Z(r) = \frac{2\pi\rho}{b} \frac{uJ_1(u)}{1 - J_o(u_a)} \tag{9.1}$$

For the impedance $r = 0$ to $r = a$, we obtain the impedance from the ratio of the voltage (8.71) with $u = u_a$ to the current in (8.57) to give (see note 23/1/8)

$$Z_a = \frac{2\pi\rho}{b} \frac{u_a J_1(u_a)}{1 - J_o(u_a)} \tag{9.2}$$

This impedance is obtained from the emf generated by the total flux entering the conductor divided by the total azimuthal current, Figure 9.1. By substituting for $u_a = -j^{1/2}ma$ then

$$Z_a = -j^{1/2}m\rho(2\pi a/b)\frac{J_1(u_a)}{J_o(u_a)[J_o^{-1}(u_a) - 1]} \tag{9.3}$$

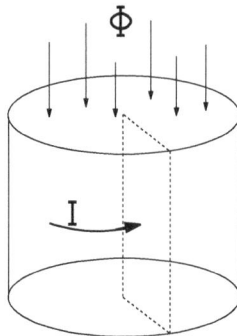

Figure 9.1 *Impedance determined from the emf generated by the flux change divided by the azimuthal current*

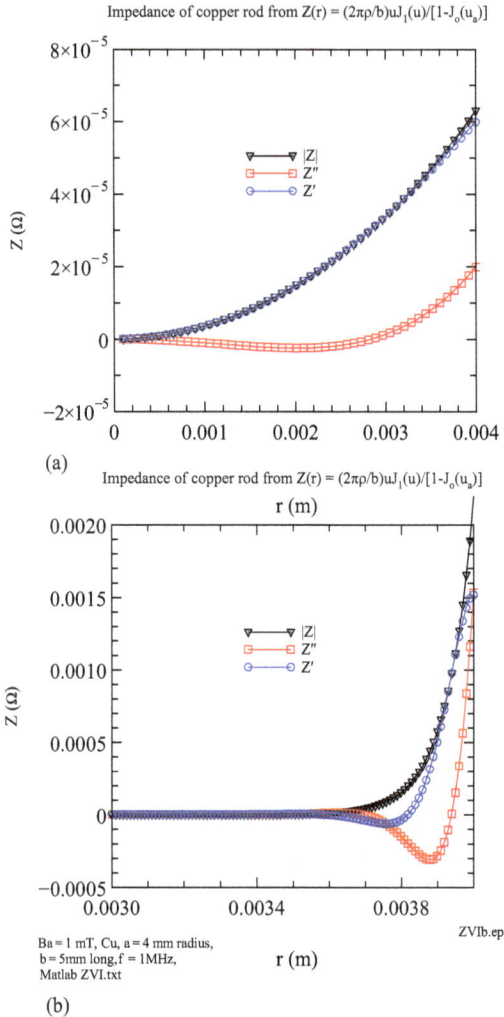

Figure 9.2   *Impedance $Z = V/I$ for a copper rod using (9.1). Applied field amplitude $B_a = 1\,mT$, copper rod radius 4 mm, length 5 mm. (a) $f = 1\,kHz$ and (b) $f = 1\,MHz$*

At low frequencies, taking only the first terms of the Bessel series we obtain

$$Z_a = \frac{2\pi\rho}{b}\frac{u_a^2/2}{(u_a/2)^2} = \frac{4\pi\rho}{b} \tag{9.4}$$

This is shown plotted in Figure 9.2.

9.1.1   Low-frequency approximations

At low frequencies such that $u_a \ll 1$ or since $u_a = maj^{3/2} = \sqrt{2}(a/\delta)j^{3/2}$, then for skin depths greater than the conductor radius a, the Bessel functions can be approximated by taking the first terms only of the series. On expanding $(u/2)^2 = (maj^{3/2}/2)^2 = \omega\mu\sigma a^2 j^3/4$, (8.57) yields the approximate current

$$I(t) = \frac{j\omega B_a \sigma a^2 b}{4} e^{j\omega t} \tag{9.5}$$

or

$$I(t) = \frac{\omega B_a \sigma a^2 b}{4}[jsin(\omega t) - cos(\omega t)] \tag{9.6}$$

This current amplitude is the same as that obtained for a conducting disc in an alternating magnetic field where the field is assumed uniform inside the disc. See references [20] or [83] and Appendix C in this chapter. Similarly, the low-frequency emf is

$$V_{emf}(t) = \omega\pi a^2 B_a[jsin(\omega t) - cos(\omega t)] \tag{9.7}$$

Also, if $u_a \ll 1$, then (9.2) becomes

$$Z_a = \frac{2\pi\rho}{b}\frac{u_a^2/2}{(u_a/2)^2} \tag{9.8}$$

and

$$Z_a = \frac{4\pi\rho}{b} \tag{9.9}$$

The low frequency power is then

$$<P> = \frac{I_o^2}{2}Re(Z) = \frac{1}{2}(\omega B_a \sigma a^2 b/4)^2\frac{4\pi\rho}{b} \tag{9.10}$$

Hence,

$$<P> = \frac{\omega^2 B_a^2 \sigma a^4 \pi b}{8} \tag{9.11}$$

The approximate (9.11) and exact power equations are shown plotted in Figure 9.3(a) and 9.3(c), respectively, where the power is normalised to $P_n = <P>/B_a^2$. Also shown plotted is the normalised power equation for induction heating a cylindrical metal sample [87], where

$$P_n = \frac{P_L}{CFB_a^2} \tag{9.12}$$

$$P_L = 2a\pi b(B_a/\mu_o)^2\sqrt{\pi\rho\mu_o\mu_r f} \tag{9.13}$$

a and b are the radius and cylinder length in (m), respectively. f is the frequency, C a coupling constant and F a transmission coefficient. These are not given in the reference but they are not necessary here because of the normalisation.

9.2 Wave impedance

The intrinsic or wave impedance is defined as follows:

$$Z_i = \frac{E_\phi}{H_z} = \frac{J_\phi}{\sigma H_z} = \rho\mu\frac{J_\phi}{B_z} \; \Omega \tag{9.14}$$

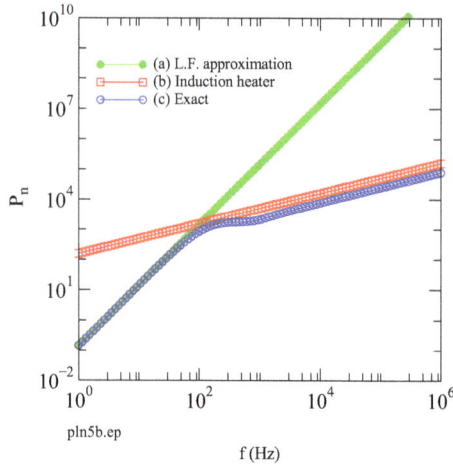

pln5b.ep

f (Hz)

Figure 9.3   Normalised power dissipated in aluminium solid cylinder, radius 12.59 mm, length 10 mm. (a) Low-frequency approximation (9.11), (b) induction heater (9.12) and (c) exact equation normalised to $P_n = \langle P \rangle / B_a^2$.

Substituting for (8.47) and (8.7) gives

$$Z_i = j^{1/2} m \rho \frac{J_1(u)}{J_o(u)} = -\rho \frac{u_a J_1(u)}{a J_o(u)}. \tag{9.15}$$

or

$$Z_i = Z_o \frac{J_1(u)}{J_o(u)} \tag{9.16}$$

where

$$Z_o = j^{1/2} m \rho = j^{1/2} (\omega \mu / \sigma)^{1/2} = (1+j)\rho/\delta \tag{9.17}$$

is the wave impedance of a semi-infinite good conductor defined by $\sigma / \omega \epsilon > 10$ [20]. This is shown plotted in Figure 9.4 and Figure 9.5.

9.2.1   Average wave impedance

The average wave impedance is defined by

$$\bar{Z}_i = \frac{2}{a^2} \int_0^a r Z(r) dr \tag{9.18}$$

If $Z(r) = Z_1$ is uniform and constant, then the average internal impedance is $\bar{Z} = (2/a^2) Z_1 \int_0^a r dr = (2/a^2) Z_1 (a^2/2) = Z_1$ as expected. Substituting (2.101) into (9.18) and converting to the variable u gives

$$\bar{Z}_i = \frac{2 j^{1/2} \rho}{a^2 m} \int_0^a u \frac{J_1(u)}{J_o(u)} du. \tag{9.19}$$

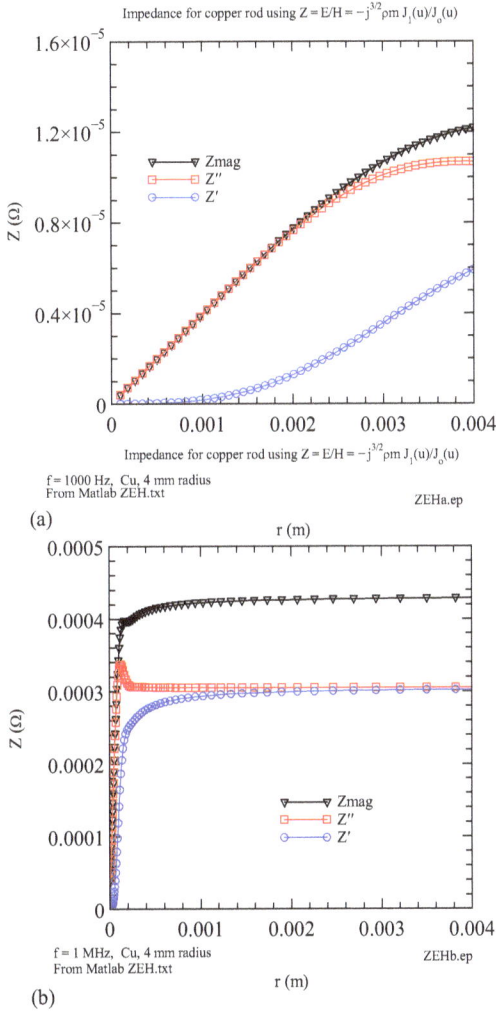

Figure 9.4   Wave impedance determined from E/H using (2.101). Applied field amplitude $B_a = 1\,mT$, copper rod radius 4 mm. (a) $f = 1\,kHz$ and (b) $f = 1\,MHz$

9.2.2   Average impedance and permeability

Substitute the permeability from (8.21) into (9.16) to give

$$Z_{\phi a} = Z_o(u_a/2) < \mu_r > \Omega \tag{9.20}$$

Substitute for $u_a = maj^{3/2}$, then

$$Z_{\phi a} = -\omega\mu_o(a/2) < \mu_r > \tag{9.21}$$

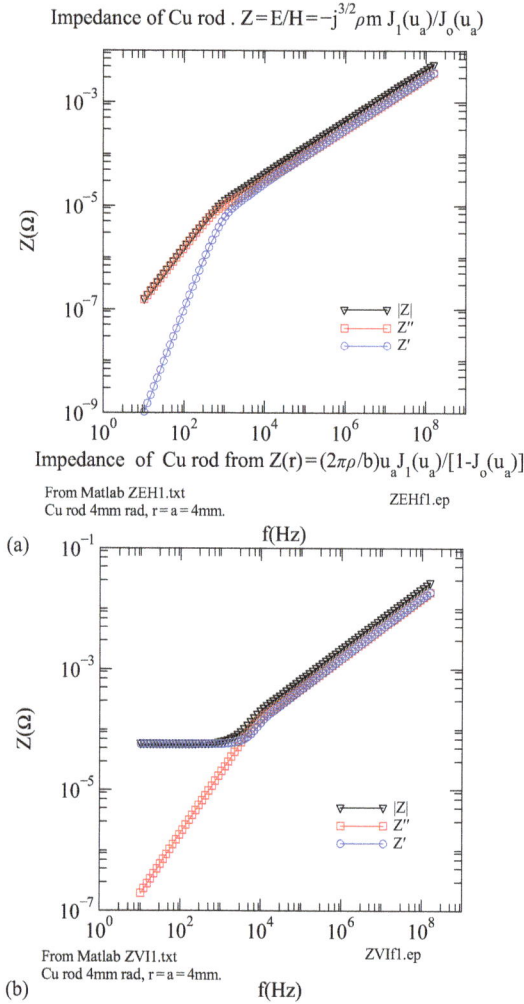

Figure 9.5   *Impedance-frequency characteristics (a) wave impedance $Z = E/H$ (9.16) and (b) impedance $Z = V/I$ (9.2). Applied field amplitude $B_a = 1\,mT$, copper rod radius 4 mm, length 5 mm, $r = a = 4\,mm$.*

or in terms of the real and imaginary components of $< \mu_r >$,

$$Z_{\phi a} = -\omega \mu_o (a/2) < \mu' - j\mu'' > \qquad (9.22)$$

Using approximate values of $< \mu_r >$ from (8.34) and (8.35)

$$\mu' = 1 + \chi' \approx 1 - \frac{1}{12}(a/\delta)^4 \qquad (9.23)$$

$$\mu'' = -\chi'' \approx \alpha^2/2 = \frac{1}{4}(a/\delta)^2 \qquad (9.24)$$

gives the quadrature components of the impedance

$$Z'_{\phi a} = \frac{\omega\mu_o a}{2}[1 - \frac{1}{12}(a/\delta)^4]$$
(9.25)

$$Z''_{\phi a} = \frac{\omega\mu_o a}{8}(a/\delta)^2$$
(9.26)

9.2.3   Power dissipation

The relationship between power flow and impedance is given by (see p. 59, (3.102)),

$$<P> = \frac{H_o^2}{2}Re[Z(\omega)] = \frac{E_o^2}{2}Re[1/Z(\omega)] \text{ Wm}^{-2}$$
(9.27)

Substitute (8.7), (8.48) and (9.22) into (9.27) gives

$$<P> = -\frac{1}{2}\left(\frac{B_a J_o(u)}{\mu_o J_o(u_a)}\right)^2 \omega\mu_o(a/2) <\mu'>$$
(9.28)

and

$$<P> = -\frac{a}{4}\left(\frac{j^{1/2}\omega B_a \rho J_1(u)}{\mu_o J_o(u_a)}\right)^2 \omega\mu_o <\mu'> \text{ Wm}^{-2}$$
(9.29)

9.3   Conducting-superconducting mixed state

9.3.1   Impedance from E/H

The impedance of the cylindrical conductor, which also contains superconducting regions, that is, the mixed state, may be considered by using the substitution $u_a = ma j^{3/2} = j\gamma a$, $(u_a/2)^2 = -(\gamma a/2)^2$, where

$$\gamma^2 = \lambda^{-2} + j(2/\delta^2).$$
(9.30)

λ is the superconducting penetration depth which in the weak field limit is given by the Gorter–Casimir equation

$$\lambda(T) = \lambda_o[1 - (T/T_c)^4]^{-1/2}$$
(9.31)

where λ_o is the penetration depth as $T \rightarrow 0$ and T_c is the critical temperature. For YBCO, $\lambda_o = 150$ nm and $T_c = 93$ K.

Substituting $u_a = j\gamma a$ in (9.15), $Z_\phi = -\rho\frac{u_a J_1(u)}{a J_o(u)}$, gives

$$Z_\phi = -j\gamma\rho\frac{J_1(u_a)}{J_o(u_a)}.$$
(9.32)

Re-write the permeability (8.21) as

$$\frac{J_1(u_a)}{J_o(u_a)} = <\mu_r>\frac{u_a}{2}$$
(9.33)

and substitute into (9.32)

$$Z_\phi = -j\gamma\rho(u_a/2) <\mu_r>.$$
(9.34)

Then, substitute for $u_a = j\gamma a$ to give

$$Z_\phi = \gamma^2 \rho (a/2) < \mu_r >. \tag{9.35}$$

Substitute for the permeability as in (9.25) and (9.41) to give

$$Z_\phi = \gamma^2 \rho (a/2) < \mu' - j\mu'' >. \tag{9.36}$$

$$Z_\phi = \frac{\rho a}{2} \left[\left(\frac{\mu''}{\lambda^2} - \frac{2\mu'}{\delta^2} \right) + j \left(\frac{\mu'}{\lambda^2} + \frac{2\mu''}{\delta^2} \right) \right] \tag{9.37}$$

We can then approximate as follows:

$$\mu' \approx 1 + \frac{1}{3}(\gamma a/2)^4 \tag{9.38}$$

$$\mu'' = -\chi'' \approx \frac{j^{-1}(\gamma a)^2}{8} \tag{9.39}$$

gives the quadrature components of the impedance

$$Z'_\phi = j\frac{\gamma^2 \rho a}{6}[3 + (\gamma a/2)^4] \tag{9.40}$$

$$Z''_\phi = \frac{\gamma^4 \rho a^3}{16} \tag{9.41}$$

9.3.2   Superconductor with mutual coupling

Assume that the supercurrent couples magnetically with the normal current. This may be modelled by an equivalent circuit of a normal inductance L_n in parallel with a superinductance L_λ, as discussed previously on p. 44, Figure 2.10. This resulted in a general equation for the propagation constant which applied to a normal conductor, superconductor or dielectric, depending on the choice of equivalent circuit. The previous analysis concerned an electric field applied to a rectangular slab. The analysis is repeated here, but for a cylindrical conductor in an applied axial magnetic field. The current density is now given by

$$\nabla \times \mathbf{H}_z = \mathbf{J}_\phi = y_T \mathbf{E} = I_\phi / (ba) \tag{9.42}$$

The specific admittance is

$$y_T = Y_T g; \quad g = d/A = 2\pi a/ba = 2\pi/b \tag{9.43}$$

where g is a geometrical constant for currents flowing around the axis of the cylinder, where d is the length of the current path, A is the cross-sectional area, a is the cylinder radius and b is the cylinder length. From Maxwell's first (3.46)

$$\nabla \times \mathbf{E}_\phi = -\mu \frac{\partial \mathbf{H}_z}{\partial t} \tag{9.44}$$

$$\nabla \times \nabla \times \mathbf{E}_\phi = -\mu \frac{\partial \nabla \times \mathbf{H}_z}{\partial t} = -\mu \frac{\partial y_T \mathbf{E}_\phi}{\partial t}. \tag{9.45}$$

Using

$$\nabla \times \nabla \times \mathbf{E} = \nabla(\nabla.\mathbf{E}) - \nabla^2 \mathbf{E} \tag{9.46}$$

and assuming $\nabla.\mathbf{E} = \rho_v/\epsilon = 0$, then

$$\nabla^2\mathbf{E}_\phi - \mu\frac{\partial y_T\mathbf{E}_\phi}{\partial t} = 0 \tag{9.47}$$

For sine waves

$$\nabla^2\mathbf{E}_\phi = j\omega\mu y_T\mathbf{E}_\phi \tag{9.48}$$

Comparing this with the general Helmholtz (2.57) for $\mathbf{E}$ gives for the propagation constant

$$\gamma^2 = j\omega\mu y_T \tag{9.49}$$

where

$$y_T = gY_T = gZ_T^{-1} = g(Z^{-1} + Z_l^{-1}) \tag{9.50}$$

where Z_l is a leakage impedance and Z is the impedance of two inductors in parallel given by Raven [47]

$$Z = \frac{Z_1Z_2 - X_M^2}{Z_1 + Z_2 - X_M^2} \tag{9.51}$$

where $X_M = j\omega M$ is the mutual inductive reactance coupling the two inductors. The mutual coupling is defined by

$$M = k\sqrt{L_1L_2} = k\sqrt{L_nL_\lambda} \tag{9.52}$$

where k is the coupling coefficient which is expected to be close to unity for a simple superconductor.

For the case of normal current coupling with a supercurrent, Figure 2.10, then

$$Z_1 = Z_n = R_n + j\omega L_n, \quad Z_2 = Z_\lambda = R_\lambda + j\omega L_\lambda, \quad X_M = j\omega M \tag{9.53}$$

where Z_n, R_n, L_n are normal impedance, resistance and inductance, respectively. $Z_\lambda, R_\lambda, L_\lambda$ are superconducting impedance, resistance and inductance, respectively. Equation (9.49), then becomes

$$\gamma^2 = j\omega\mu g(Z^{-1} + Z_l^{-1}) \tag{9.54}$$

$$\gamma = \sqrt{j\omega\mu g\left[\frac{Z_1 + Z_2 - X_M^2}{Z_1Z_2 - X_M^2} + Z_l^{-1}\right]} \tag{9.55}$$

To check this put $X_M = 0$, $Z_2 = \infty$, $Z_1 = R_n$, $1/Z_l = j\omega C_n$. Hence,

$$\gamma^2 = j\omega\mu g(j\omega C_n + 1/R_n) = -\omega^2\mu gC_n + j\omega\mu g/R_n = -\omega^2\mu\epsilon + j\omega\mu\sigma \tag{9.56}$$

This agrees with (2.58). Note that if ϵ or μ is a complex, then the real and imaginary parts of (9.55) are modified. The azimuthal impedance of a type II superconducting rod in an axial magnetic field is then given by (9.15), restated here as

$$Z_\phi = -\rho\frac{u_aJ_1(u_a)}{aJ_o(u_a)} \tag{9.57}$$

where $u_a = j\gamma a$ and γ from (9.55). These results may be compared with the impedance determined from V/I (8.58,8.71) and compared with known values of impedance determined experimentally, [56,84–86f]. These results expressed by (9.57) apply to any conductor, dielectric or superconductor, depending upon the choice of the equivalent circuit parameters.

9.4   Theoretical results

Figure 9.6 (MU12g2.ep) shows plots of μ' and μ'' versus δ/a using (8.25) and 1, 2 and 5 terms from the Bessel series. Note agreement between 2 and 5 terms for $\delta/a \geq 1.5$. The data was obtained using BASIC programmes MU12g2.BAS, which produced data file MU12g2.DAT. This was plotted using Easyplot, which produced plot MU12g2.EP. All files are contained in the root directory AAbes and subdirectories BASIC, Figs and Tex.

Figure 9.7 (MU12gb.ep) shows plots of μ' and μ'' versus ρ/ρ_1, where $\rho_1 = \omega\mu_o a^2/2$. If ρ_1 and μ are known then the sample resistivity can be determined from this universal graph.

9.5   Experimental results

9.5.1   Resistivity measurements

The resistivity of a copper rod, dimensions 30 mm long and 4 mm in diameter, was measured by passing direct currents of up to 50 A through the rod and measuring

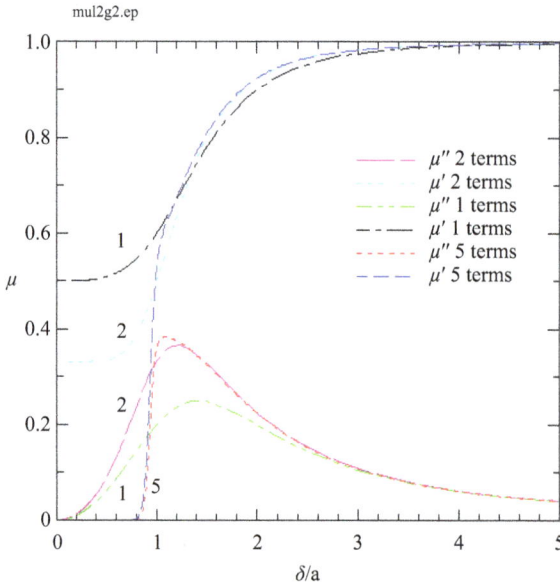

Figure 9.6   Complex permeability dependence on skin depth/rod radius from Bessel functions(mu12g2.eps)

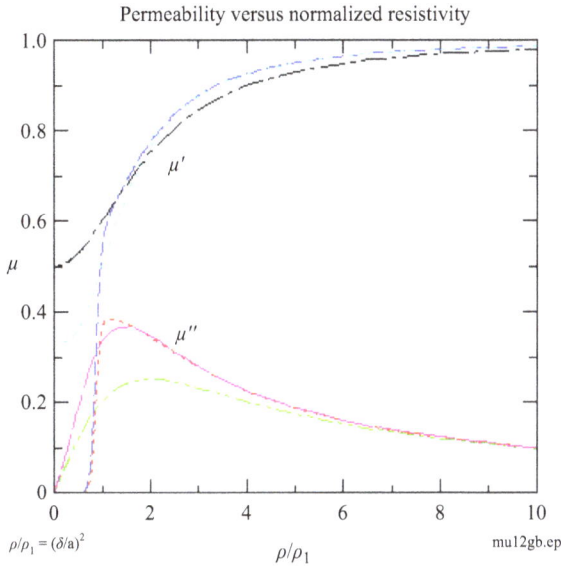

Permeability versus normalized resistivity

*Figure 9.7   Complex permeability dependence on resistivity from Bessel
functions(mu12gb.eps)*

the voltage drop using four-terminal techniques and a high-impedance digital voltmeter. The current was then reversed, and the mean value obtained. The I/V results
are shown in Figure 9.8 for measurements at room temperature and 77 K. The mean
distance between the voltage contacts was $\bar{l} = 8.81$ mm. This yielded resistivities of
$\rho = 2.81$ nΩ m at 77 K and $\rho = 23.5$ nΩ m $= 2.35$ $\mu\Omega$ cm at 300 K for this particular
sample.

9.5.2   Susceptibility measurements

The susceptibility of the above copper rod was measured using a differential ac susceptometer [81,82]. The applied field amplitude ranged up to about 25 mT at several
different frequencies. All measurements were carried out at a temperature of 77 K.
Figure 9.9 (curoda1b.ep) shows complex differential voltages V'_D, V''_D obtained from
the copper rod described above at frequencies of 20, 50 and 100 Hz.

The susceptibility was calculated from the differential voltage measurements
using

$$\chi'_{ext} = \frac{V'_D}{V_{Do}}, \quad \chi''_{ext} = \frac{V''_D}{V_{Do}} \tag{9.58}$$

where $V_{Do} = \alpha(N_s/L_s)v_m\omega B_o$, $\alpha = 0.5498$, $N_s = 2500$, $L_s = 0.013$ m, $v_m = \pi r^2 l$,
$l = 0.03$ m, $r = 0.002$ m. Putting $V_{Do} = uB_o$ gives $u(20\,\text{Hz}) = 5.022$, $u(50\,\text{Hz}) =$
12.5566 and $u(100\,\text{Hz}) = 25.113$. Figure 9.10(curada2.ep) shows plots of χ_{ext}
obtained directly from Figure 9.9 using (9.58) and the above values of V_{Do}. The mean
values obtained are listed in Table 9.1.

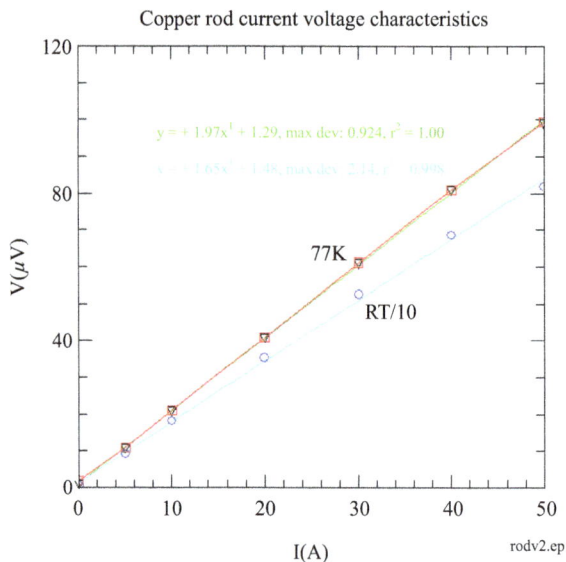

Figure 9.8   *Current voltage characteristics for a copper rod at 77 K and room temperature (RT/10)(rodv2.eps)*

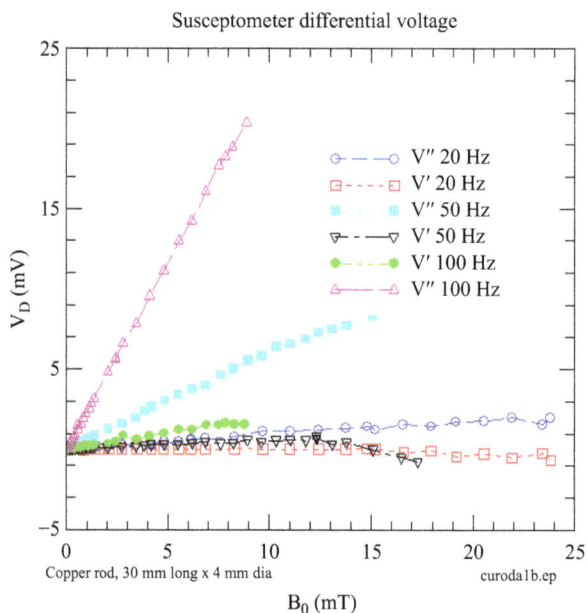

Figure 9.9   *(Measured complex differential voltage V_D at 77 K as a function of applied magnetic field B_o for several frequencies, determined using a magnetic susceptometer(tcua1b2.eps)*

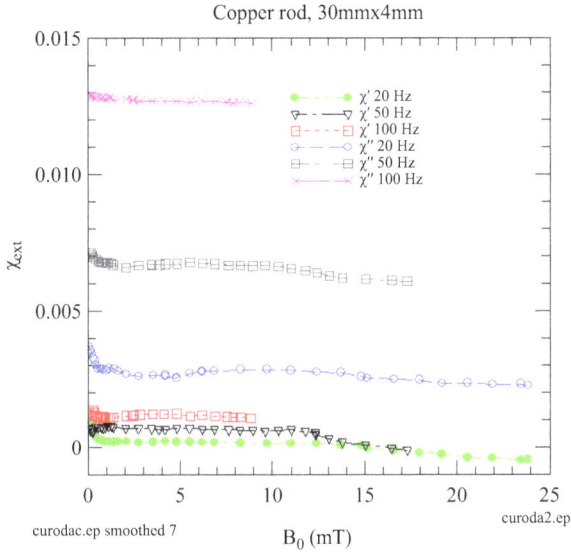

Copper rod, 30mmx4mm

curodac.ep smoothed 7

curoda2.ep

Figure 9.10 Complex susceptibility determined directly from the differential voltage measurements in Figure 9.9(curoda2.eps)

Table 9.1 Comparison of theoretical and measured mass susceptibilities of copper at 77 K. Mass susceptibility $\chi_m = (\mu_r - 1)/\rho_{Cu}$ (MUCU0B.BAS).

f Hz	χ'_{ext} $\times 10^{-4}$	χ''_{ext} $\times 10^{-3}$	$-\chi'_{th}$ $\times 10^{-8}$	χ''_{th} $\times 10^{-6}$	$-\chi'_{exp}$ $\times 10^{-8}$	χ''_{exp} $\times 10^{-6}$	ρ' nΩ m
20	1.84	2.78	11.8	3.15	2.059	0.307	6.75
50	5.52	6.65	73.6	7.82	6.177	0.744	9.76
100	11.45	12.77	287	15.3	12.822	1.43	13.57

Theoretical and experimental ac mass susceptibilities for copper are listed in Table 9.1. The theoretical values were determined from five terms of the Bessel series. At these frequencies, only a small error is introduced using two terms of the Bessel series or (8.34) and (8.35). Also, the error due to demagnetisation effects is negligible. In these calculations, the specimen dimensions were those used in the experiment, and the resistivity corresponded to the measured dc value of 2.81 nΩ m. at 77 K. The literature value of the mass susceptibility at 100 K is $\chi_m = -0.113 \times 10^{-8} m^{-3}$. The final column in this table shows the resistivity calculated from (8.36) using measured χ values. This result is also shown plotted in Figure 9.11 (LGf2.ep).

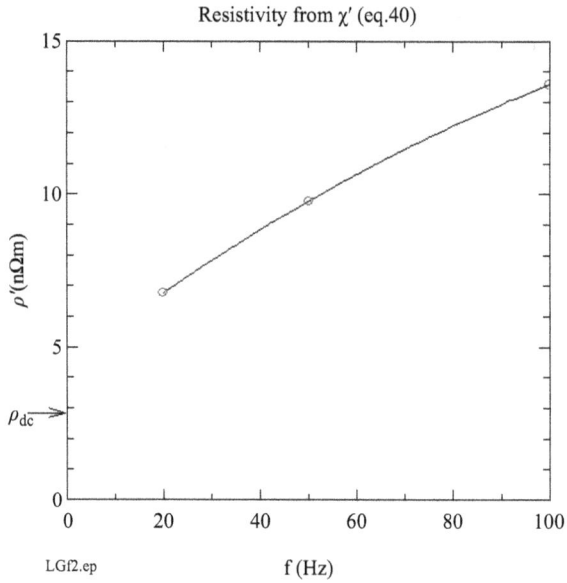

Figure 9.11   Resistivity calculated from (8.36) using measured χ values(LGf2.eps)

Chapter 10
Hollow cylindrical conductors

10.1 Introduction

In this chapter and the following chapter, the dc and ac properties of copper tubes are examined, respectively. Tubular conductors are important in many areas of electrical engineering and physics, ranging from power frequencies to microwaves. Copper or aluminium tubes are also widely used as busbars in HV substations [101]. Compared with solid busbars, they are apparently lighter and cheaper to manufacture [102]. These busbars must be able to withstand very high transient currents and voltage switching transients. The recent disaster at the Hayes substation, which caught fire and shut down Heathrow airport for a day, indicates the importance of reliable busbars [103]. However, the exact cause of this disaster has not yet been discovered.

10.2 Cu tube resistivity, density and – axial dc measurements

In the following, we have analysed standard Cu–DHP copper tubes (deoxidised oxygen-free copper) [100]. Firstly, dc properties and then ac properties are presented. For steady state dc measurements on tubes, the analysis and measurements are relatively simple compared with the ac case, since for dc, the tube geometry and resistivity determine the resistance. For measurements on Cu–DHP copper tubes, the resistivity of the copper is higher than pure copper due to the presence of phosphorus in the manufacture of the tubes, [100]. The phosphorus, at between 0.015% and 0.04%, is included to improve the mechanical performance of the tubes. However, this raises the resistivity of the copper from $17\,\mathrm{n\Omega m}$ for pure copper to about $21.7\,\mathrm{n\Omega m}$ for Cu–DHP copper.

In the dc case the measured resistivity is obtained from

$$\rho = \frac{A_t R_t}{l_e} \tag{10.1}$$

where R_t is the tube resistance, l_e is the electrical length between the voltage contacts. A_t is the tube cross sectional area. This is simply given by

$$A_t = \pi(r_{od}^2 - r_{id}^2) \tag{10.2}$$

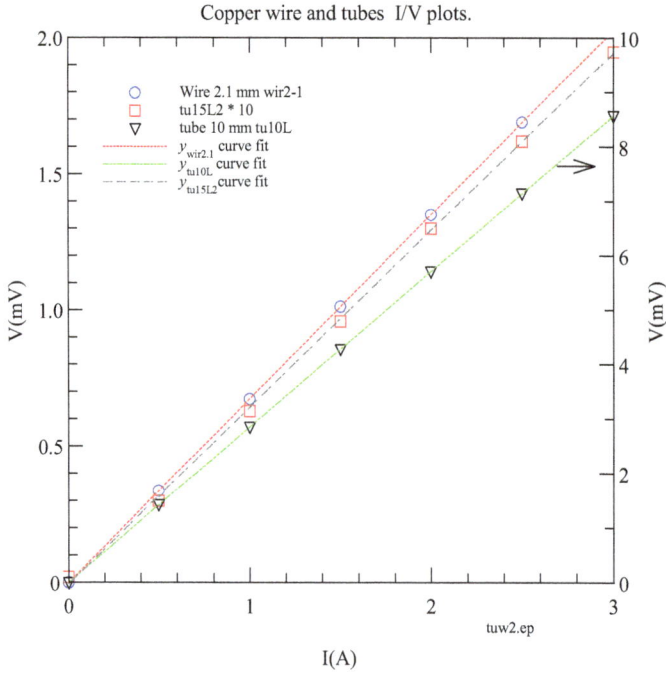

Figure 10.1   *DC I/V measurements for copper wire wir2.1 and tubes tu10L, tu15L2. See Table 10.1*

where r_{od} and r_{id} are the outer and inner tube radius, respectively. Alternatively, if the thickness of the tube wall t_w is measured then A_t is given approximately by

$$A_t = \pi(r_{od} - t_w/2)^2 \tag{10.3}$$

The resistivity was measured using dc four terminal I/V techniques. The high and low currents HI, LI, were injected and extracted to the ends of the tube, respectively. The high and low voltages HV, IV were measured a few cm from the current terminals. The current and voltage terminals were made either by soldering wires to the copper tube or using copper wire rings screwed tightly to the tubes. Constant currents up to 3 A were supplied from a programmable power supply, Tenma 72-2720. The voltages measured using a high-input impedance digital voltmeter, Solartron 7150+, range 0.2 V, 6.5 digit sensitivity 100 nV, impedance $\geq 10\,G\Omega$. The I/V results for several specimens are shown in Figure 10.1. The curve fits for the data are given by

$$\begin{aligned} V_{wir2.1} &= 0.676I - 8.10 \times 10^{-4}, \quad maxdevn : 0.00130 \\ V_{tu10L} &= 2.86I + 6.79 \times 10^{-4}, \quad maxdevn : 0.00421 \\ V_{tu15L2} &= 0.0650I - 6.43 \times 10^{-4}, \quad maxdevn : 0.0264 \end{aligned} \tag{10.4}$$

This method of determining the tube cross sectional area (10.2) gave good results for solid copper wire, Table 10.1.

Table 10.1   Cu tube resistivity and density measurements

Sample	Resistivity $n\Omega m$	Error %	Density kgm^{-3}	Error %
Cu	17	0	8930	0
Cu–DHP	21.74	0	8900	0
Wire 2.1 mm	17.2	1	9078	1.6
tu10L	20.58	−5	8495	−4.5
tu10Ls1	19.5	−10	8837	−1
tu10Ls2	20.6	−5	8436	−5
tu15L2	23.34	7	9130	2
ring28	29.34	10	9044	1.3

However, for copper tubes, this gave erroneous results. Possibly because the radius varied slightly along the length of the tube, making a significant variation in the area A_t.

A superior method was to immerse these samples in water and measure the volume of water displaced, v_d, as described below. Using this displacement measurement technique, the copper tube agreed to within a few per cent of the density of Cu–DHP, Table 10.1. The tube area is then given by

$$A = \frac{v_d}{l_t} \tag{10.5}$$

where l_t is the tube length, from this the resistivity was obtained from (10.1). The values obtained also agreed with the Cu–DHP resistivity within a few percent, Table 10.1.

In Table 10.1 the copper wire was 2.1 mm diameter, total length 22.4 cm, weight 7.042 g. This gave a density of 9078 kg m^{-3}. The I/V measurements gave an average resistance of 0.676 mΩ for an electrical length 13.6 cm long. The cross-sectional area of the wire was 3.463 mm^2. This gave a resistivity of $\rho = 17.2$ nΩ m.

The sample tu10L refers to a hollow copper tube with a diameter of 10 mm, a wall thickness of 0.7 mm, a total length of 3.04 m, an electrical length of 2.84 m, bent into a loop (*U* shape) with gap of 10 mm between plastic bars, Figure 10.3. The following details were etched on the tube: "BS1057 OUTOKUMPU tube-e. Premium 10x0.7 WWW tube-e.com".

The distance between voltage contacts was 284 cm. The I/V measurements, Figure 10.2, yielded and average resistance of 2.859 mΩ and resistivity $\rho = 20.58$ nΩm. The tube weight was 528.1 g. This gave a density of 8495 kgm^{-3}.

Specimens tu10Ls1, Ls2 were copper strips taken from tu10L ends.

Specimens tu15L2 was a copper tube 15.87 mm average outer diameter, average wall thickness 0.946 mm total length 16.7 cm, electrical length between voltage contacts 11.5 cm. The I/V measurements yielded an average resistance $R = 65\ \mu\Omega$, resistivity 23.34 nΩm.

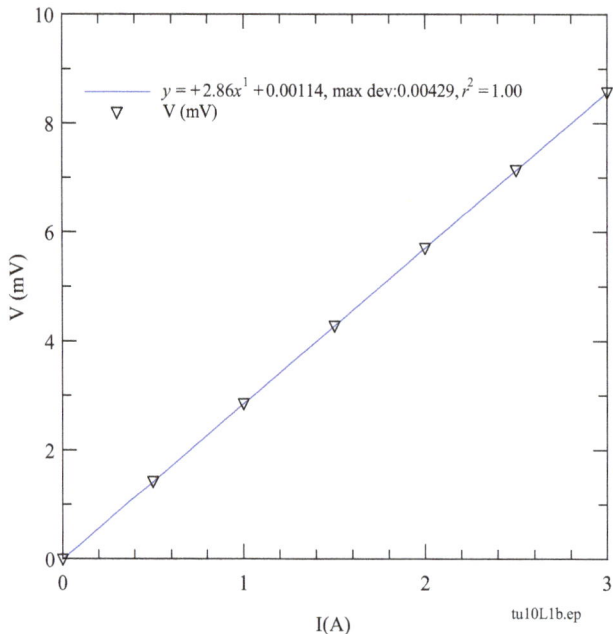

Figure 10.2   Cu tube, 304 cm long 10 mm diameter. DC I/V measurements. Yields $R_{ave} = 2.86\,m\Omega$. This gives resistivity $\rho = 20.58\,n\Omega m$. See Table 10.1.

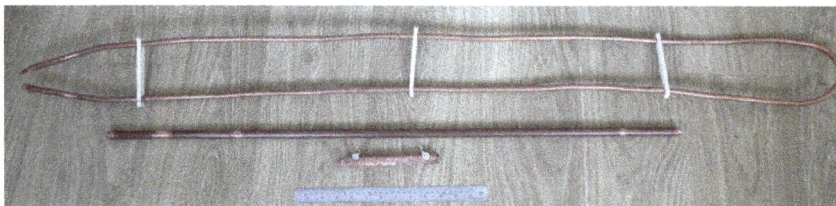

Figure 10.3   Specimens top to bottom: tu10L, tu15L, and tu15L2. Refer to Table 10.1

10.2.1   Measurement of sample volume by water displacement – Archimedes principle [99]

The sample was placed in a plastic tube sealed at the bottom (Figure 10.4). A mark was made on the tube wall m_1 where the maximum height of the sample occurred. The sample was then removed, and water was inserted into the mark m_1 on the wall. The sample was then re-inserted and a second mark m_2 made where the water reached. The displacement length was then given by $m_2 - m_1$.

For example, sample tu15L2. The increase in water volume after inserting the Cu tube was 6.9 ml as measured with a syringe. The tube weight was 63 g. Hence the

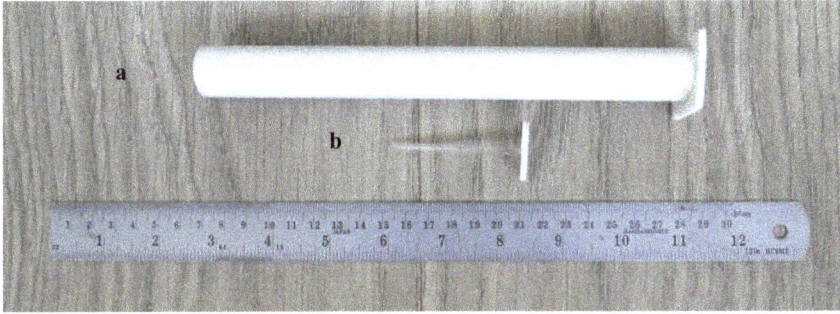

*Figure 10.4   Containers for measuring specimen volumes by the water
displacement method (a) large specimens and (b) small specimens*

mass density was $63/6.9 = 9.13\,gcm^{-3}$ or $9130\,kgm^{-3}$. This is 2% higher than that given for Cu–DHP, see Table 10.1.

10.3   Inductance of a hollow tube and solid cylindrical conductor

Equations for the inductance of a solid wire are well known [20]. Equations for the inductance of tubular conductors are also available [106]. However, James Clerk Maxwell was one of the first to derive expressions for these conductors, and we briefly give details of his results.

In the theory due to Maxwell [3] for the inductance of tubes, we assume a very long loop with forward and return tubes and consider only a small section of the loops far from the ends. The length of the tube is considered l, the internal radius is a_2, the outside radius is a_1 and the distance between the forward and return tube axes, b. Using magnetic induction and the kinetic energy for this arrangement, Maxwell obtained the inductance as follows:

$$L/l = 2\mu_e \ln \frac{b^2}{a_1 a_1\prime} + \frac{\mu_i}{2} \left[\frac{a_1^2 - 3a_2^2}{a_1^2 - a_2^2} + \frac{4a_2^4}{a_1^2 - a_2^2} \ln \frac{a_1}{a_2} \right] + [``(\prime)] \qquad (10.6)$$

where μ_e is the relative permeability outside the conductor and μ_i the value inside the conductor. The equation in the second square brackets applies to the return tube. This equation is the same as the first except that all terms are replaced by primes ($\prime$). If the conductors are solid wires, a_2 and a_2' are zero then

$$L/l = 2\mu_e \ln \frac{b^2}{a_1 a_1'} + \frac{\mu_i + \mu_i'}{2} \qquad (10.7)$$

Converting to SI units by putting $\mu_e = \mu_e/4\pi$ and $\mu_i = \mu_i/4\pi$ we obtain

$$L/l = \frac{\mu_e}{2\pi} \ln \frac{b^2}{a_1 a_1'} + \frac{\mu_i}{8\pi} \left[\frac{a_1^2 - 3a_2^2}{a_1^2 - a_2^2} + \frac{4a_2^4}{a_1^2 - a_2^2} \ln \frac{a_1}{a_2} \right] + [``(\prime)] \qquad (10.8)$$

and for a solid wire

$$L/l = \frac{\mu_e}{2\pi} \ln \frac{b^2}{a_1 a_1'} + \frac{\mu_i + \mu_i'}{8\pi} \tag{10.9}$$

If $a_1 = a_1'$, $a_2 = a_2'$, and $\mu_i = \mu_i'$ the tube inductance is then

$$L/l = \frac{\mu_e}{\pi} \ln \frac{b}{a_1} + \frac{\mu_i}{4\pi} \left[\frac{a_1^2 - 3a_2^2}{a_1^2 - a_2^2} + \frac{4a_2^4}{a_1^2 - a_2^2} \ln \frac{a_1}{a_2} \right] \tag{10.10}$$

and the inductance of a solid wire is

$$L/l = \frac{\mu_e}{\pi} \ln \frac{b}{a_1} + \frac{\mu_i}{4\pi} \tag{10.11}$$

As mentioned, an expression similar to (10.10) was obtained using flux linkages in a tubular conductor [106]. Equation (10.11) for a solid wire has also been derived using vector potentials [20].

Maxwell points out that "If the wires are magnetic, the magnetism in them will disturb the magnetic field and we cannot apply the preceding reasoning. Equations (10.6) and (10.7) are only strictly true for $\mu_i = \mu_i' = \mu_e$". However, ferromagnetic conductors can be considered, provided they are not magnetised, but this may be difficult to avoid.

Chapter 11

Hollow cylindrical conductor – axial AC: experimental measurements

11.1 Introduction

In this chapter, we present some experimental impedance measurement results obtained from hollow cylindrical conductors and compare these with theory. The experimental techniques employed include Hewlett-Packard Gain-Phase Meter (GPM) HP3575A, Racal LCR Meter 9343M, Wayne Kerr Universal Bridge B224, and HP Impedance Analyzer HP4192A. All the instruments were 4-terminal techniques except the GPM, which was two-terminal. The four terminal techniques reduce parasitic external error terms. In the GPM case, a process of de-embedding was used, which involved cancelling out parasitic external error terms [104]. The GPM technique had the significant advantage of permitting the application of higher drive currents.

11.2 Tube impedance theory

A widely used equation for determining the internal impedance of a tubular conductor, internal radius q, and outside radius r, is given by [106]

$$Z = \rho m/(2\pi qD)[I_0(mq)K_1(mr) + I_1(mr)K_0(mq)] \tag{11.1}$$

where

$$D = [I_1(mr)K_1(mq) - I_1(mq)K_0(mr)] \tag{11.2}$$

In these equations, I and K are modified Bessel functions of the first kind, ρ is the resistivity of the conductor and

$$m = \sqrt{j\omega\mu\sigma} \tag{11.3}$$

The theoretical results using equation 11.1 are shown plotted in Figure 11.1.

11.3 Experimental results

Figure 11.2 shows impedance results obtained from sample tu10L previously described in chapter 10 and Table 10.1. Experiments from GPM measurements [1,104] and theory using (11.1).

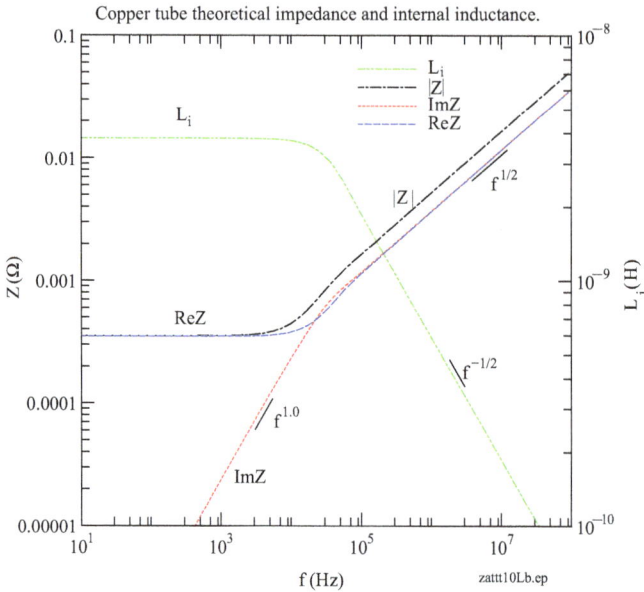

Figure 11.1   Frequency dependance of impedance and inductance for a hollow copper tube showing skin effect. See Chapter 11, (11.1). In this example the tube length was 3.04 m with inner radius 4.3 mm and outer radius 5 mm.

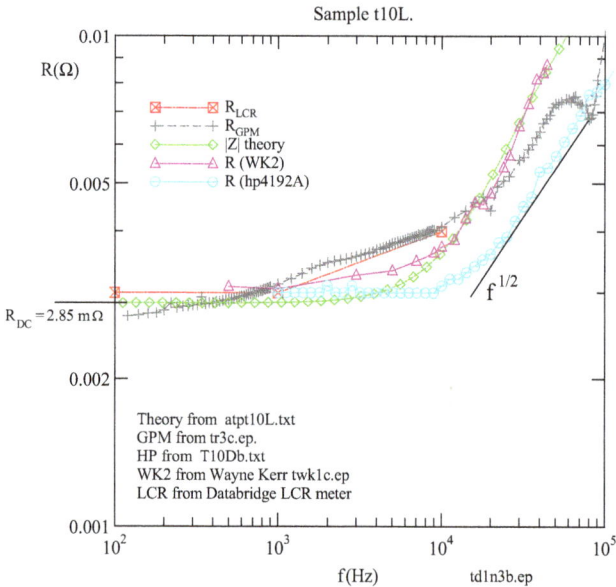

Figure 11.2   AC resistance of a long hollow Cu-DHP tube, 10 mm dia 3.04 m long, sample tu10L. Theory from (11.1). Experimental measurements using gain phase meter (GPM) HP3575A, Impedance Analyzer HP4192A, AC Bridge Wayne Kerr B224, LCR meter databridge 9343M.

Experimental inductance measurements
of 10 cm dia Cu Tube tu10L.

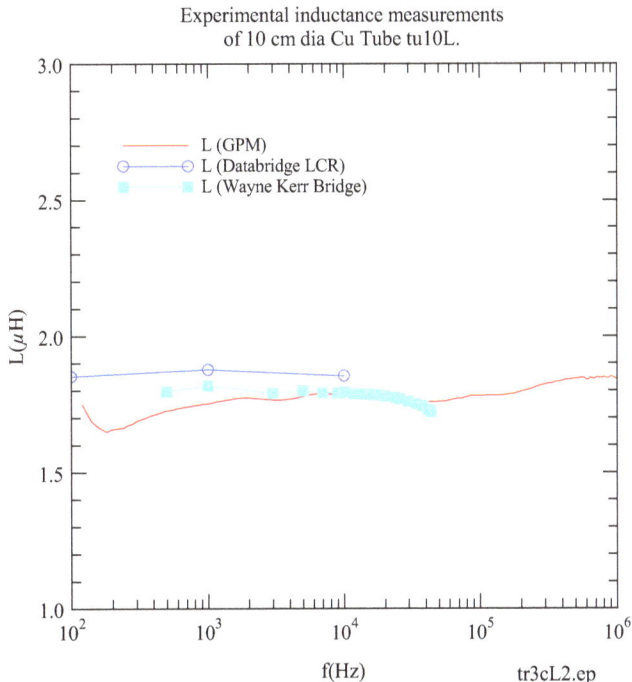

Figure 11.3   Experimental inductance measurements of 10 cm dia Cu Tube tu10L. GPM, [104], with PA and Rs = 3.305 Ω, 35 nH. FG = 1 Vrms. Wayne Kerr Bridge B224, Racal LCR Databridge 9343M.

Figure 11.3 shows inductance results obtained from the same sample tu10L. The average inductance is about 1.8 μH.

The dc inductance obtained from (10.10) gives a total value of 1.11 μH. This is considerably lower than the measured ac value. However, the calculated value of the external value is Le = 1.8214 μH, which is closer to the measured value of 1.8 μH.

The dc inductance was calculated for sample tu10L with dimensions in m: $l = 1.52$, $a1 = 5.1e - 3$, $a2 = 4.31e - 3$, and $b = 0.1$. This gives for the inductances in μH: external Le = 1.8214, internal Li = -0.7114 and total $L = 1.11$.

Chapter 12

Hollow cylindrical
conductor – impedance analysis

Much of the analysis on solid cylindrical conductors can be modified and applied to the case of a long hollow cylindrical conductor or tube. DC analysis and testing, along with a simplified ac analysis based on one skin depth, have already been considered. In this chapter, we consider the detailed analysis based on Bessel functions [115]. Conducting tubes excited by an axial magnetic field are considered further on.

12.1 Current density

As for solid cylinders, we consider an ac generator driving a constant current $I = I_o e^{j\omega t}$ parallel to the z-axis of the tube. The tube has outside radius a, inside radius b, and length l. The following analysis is initially similar to the solid cylindrical conductor case. Still, in this case, for tubular conductors, the Bessel function solutions need to include Bessel functions of the second kind.

For a good conductor with current flowing in the z-direction, and assuming the displacement current is zero, then

$$J_z = \sigma E_z, \quad E_z = \rho J_z \tag{12.1}$$

Equation (5.9) becomes

$$\frac{\partial^2 J_z}{\partial r^2} + \frac{1}{r}\frac{\partial J_z}{\partial r} - \mu\sigma\frac{\partial J_z}{\partial t} = 0. \tag{12.2}$$

12.1.1 Sine waves

For sine waves, $J_z(r,t) = J_z e^{j\omega t}$ and $\partial/\partial t = j\omega$, (12.2) becomes

$$\frac{\partial^2 J_z}{\partial r^2} + \frac{1}{r}\frac{\partial J_z}{\partial r} - j\omega\mu\sigma J_z = 0 \tag{12.3}$$

Putting

$$u = mrj^{3/2}, \quad m = \sqrt{(\omega\mu\sigma)} \tag{12.4}$$

$$\frac{\partial^2 J_z}{\partial u^2} + \frac{1}{u}\frac{\partial J_z}{\partial u} + J_z = 0 \tag{12.5}$$

Hence J and E fields are given by a zero order Bessel equation with complex argument u, where $u = mrj^{3/2}$. The general solution is given by

$$J_{zr} = AJ_o(u) + BN_o(u) \tag{12.6}$$

where $J_o(u)$ and $N_o(u)$ are zero order Bessel functions of the first and second kind, respectively. (See the general Appendix.)

For the outer surface of the cylinder $r = b$, the solution is

$$J_{zb} = AJ_o(u_b) + BN_o(u_b) \tag{12.7}$$

For the inner surface of the cylinder $r = a$, the solution is

$$J_{za} = AJ_o(u_a) + BN_o(u_a) \tag{12.8}$$

where both J_{za} and J_{zb} are constants for a given frequency. Solving (12.7) and (12.8) for A and B, respectively.

$$A = [J_{zb}N_o(u_a) - J_{za}N_o(u_b)]/D \tag{12.9}$$

$$B = [J_{za}J_o(u_b) - J_{zb}J_o(u_a)]/D \tag{12.10}$$

$$D = J_o(u_b)N_o(u_a) - J_o(u_a)N_o(u_b) \tag{12.11}$$

Although we know J_{zb} because this is the applied current. J_{za} is unknown but can be obtained by differentiating J_{zb} and J_{za} wrt u_b and u_a as indicated by the prime (').

$$J'_{zb} = 0 = AJ'_o(u_b) + BN'_o(u_b) \tag{12.12}$$

$$J'_{za} = 0 = AJ'_o(u_a) + BN'_o(u_a) \tag{12.13}$$

Both J'_{za} and J'_{zb} are zero because these fields are constants at the surfaces a and b. Hence,

$$AJ'_o(u_b) + BN'_o(u_b) = AJ'_o(u_a) + BN'_o(u_a) \tag{12.14}$$

$$\frac{A}{B} = \frac{N'_o(u_a) - N'_o(u_b)}{J'_o(u_b) - J'_o(u_a)} \tag{12.15}$$

Abbreviate the parameters
$J_o(u_r) = J_r, \; J_o(u_a) = J_a, \; J_o(u_b) = J_b$
$N_o(u_r) = N_r, \; N_o(u_a) = N_a, \; N_o(u_b) = N_b$
$J_z(u_b) = J_{zb}, \; J_z(u_a) = J_{za}, \; J_z(u_r) = J_{zr}$

For (12.76) and (12.7), substitute for A and B from (12.9), (12.10), and (12.11)

$$J_{zr} = \frac{J_{zb}N_a - J_{za}N_b}{D}J_r + \frac{J_{za}J_b - J_{zb}J_a}{D}N_r \tag{12.16}$$

$$J_{zb} = \frac{J_{zb}N_a - J_{za}N_b}{D}J_b + \frac{J_{za}J_b - J_{zb}J_a}{D}N_b \tag{12.17}$$

where

$$D = J_bN_a - J_aN_b \tag{12.18}$$

From (12.16) and (12.17) put

$$c_1 = a_1x + b_1y \tag{12.19}$$

$$c_2 = a_2x + b_2y \tag{12.20}$$

where

$$c_1 = J_{zb}, a_1 = J_b, x = A, b_1 = N_b, y = B, c_2 = J_{zr}, a_2 = J_r,$$
$$x = A, b_2 = N_r, y = B \tag{12.21}$$

Solving (12.19) and (12.20) for x and y

$$x = \frac{c_1b_2 - c_2b_1}{a_1b_2 - a_2b_1} \tag{12.22}$$

$$y = \frac{a_1c_2 - a_2c_1}{a_1b_2 - a_2b_1} \tag{12.23}$$

Substituting for the parameters from (12.21) gives

$$A = \frac{J_{zb}N_r - J_{zr}N_b}{J_bN_r - J_rN_b} \tag{12.24}$$

$$B = \frac{J_{zr}J_b - J_{zb}J_r}{J_bN_r - J_rN_b} \tag{12.25}$$

Hence,

$$\frac{A}{B} = \frac{J_{zb}N_r - J_{zr}N_b}{J_{zr}J_b - J_{zb}J_r} = \frac{N_a' - N_b'}{J_b' - J_a'} \tag{12.26}$$

Solving for J_{zr} gives

$$J_{zr} = J_{zb}\frac{N_rK_1 - J_rK_2}{N_bK_1 - J_bK_2} \tag{12.27}$$

where $K_1 = J_a' - J_b'$, $K_2 = N_a' - N_b'$ are constants for fixed frequency.
Obtaining the surface current density at a from (12.27), where N_r becomes N_a and J_r becomes J_a gives

$$J_{za} = J_{zb}\frac{N_aK_1 - J_aK_2}{N_bK_1 - J_bK_2} \tag{12.28}$$

Inserting the full parameters gives

$$J_z(u_r) = J_z(u_b)\frac{N_o(u_r)K_1 - J_o(u_r)K_2}{N_o(u_b)K_1 - J_o(u_b)K_2} \tag{12.29}$$

$$J_z(u_a) = J_z(u_b)\frac{N_o(u_a)K_1 - J_o(u_a)K_2}{N_o(u_b)K_1 - J_o(u_b)K_2} \tag{12.30}$$

where $K_1 = J_o'(u_a) - J_o'(u_b)$, $K_2 = N_o'(u_a) - N_o'(u_b)$ are constants for fixed frequency.

12.1.2   Wave impedance

The wave impedance is defined by

$$Z_z = \frac{E_z}{H_\phi} \tag{12.31}$$

where E_z is given by

$$E_z = \rho J_{zr} \tag{12.32}$$

ρ is the resistivity of the metal tube. J_{zr} is given by (12.77) and H_ϕ by (5.97).

$$H_\phi(r) = \frac{1}{j\omega\mu}\frac{dE_z}{dr} \tag{12.33}$$

$$H_\phi(u) = mj^{-1/2}\frac{dE_z}{du} \tag{12.34}$$

From (12.32)

$$H_\phi = m\rho j^{-1/2}\frac{dJ_{zr}}{du} \tag{12.35}$$

The impedance at r then becomes

$$Z_z(u_r) = \frac{E_z}{H_\phi} = m\rho j^{-1/2}\frac{J_z(u_r)}{J_z'(u_r)}, \ \Omega \tag{12.36}$$

Inserting the full parameters gives

$$Z_z(u_r) = m\rho j^{-1/2}\frac{N_o(u_r)K_1 - J_o(u_r)K_2}{N_o'(u_r)K_1 - J_o'(u_r)K_2}, \ \Omega \tag{12.37}$$

Since

$$J_o'(u) = -J_1(u), \ N_o'(u) = -N_1(u) \tag{12.38}$$

$$Z_z(u_r) = -m\rho j^{-1/2}\frac{N_o(u_r)[J_1(u_a) - J_1(u_b)] - J_o(u_r)[N_1(u_a) - N_1(u_b)]}{N_1(u_r)[J_1(u_a) - J_1(u_b)] - J_1(u_r)[N_1(u_a) - N_1(u_b)]}, \ \Omega \tag{12.39}$$

where $K_1 = -[J_1(u_a) - J_1(u_b)], K_2 = -[N_1(u_a) - N_1(u_b)]$.

12.2   Current amplitude

The current supplied to the tube is given by

$$I = I_o e^{j\omega t} \tag{12.40}$$

where I_o is the amplitude.

$$I_o = \int_a^b J_z(r)ds = \int_a^b J_z(r)2\pi r dr \tag{12.41}$$

From (12.4)

$$u = mrj^{3/2}, \ r = u/(mj^{3/2}), \ dr = du/(mj^{3/2}), \ rdr = udu/(m^2 j^3) \tag{12.42}$$

Hence

$$I_o = \frac{2\pi}{m^2 j^3} \int_a^b J_z(u) u du \tag{12.43}$$

From (12.77)

$$J_z(u_r) = J_z(u_b) \frac{N_o(u_r)K_1 - J_o(u_r)K_2}{N_o(u_b)K_1 - J_o(u_b)K_2} \tag{12.44}$$

where $K_1 = J_o'(u_a) - J_o'(u_b)$, $K_2 = N_o'(u_a) - N_o'(u_b)$ are constants for fixed frequency.

Substituting for $J_z(u_r)$ gives

$$I_o = \frac{2\pi K_3}{m^2 j^3} \int_a^b [N_o(u_r)K_1 - J_o(u_r)K_2] u du \tag{12.45}$$

where

$$K_3 = J_z(u_b)/[N_o(u_b)K_1 - J_o(u_b)K_2] \tag{12.46}$$

The Bessel function integrals are given by

$$\int u J_o(u) du = u J_1(u), \quad \int u N_o(u) du = u N_1(u) \tag{12.47}$$

$$I_o = \frac{2\pi K_3}{m^2 j^3} [u N_1(u_r)K_1 - u J_1(u_r)K_2]_a^b \tag{12.48}$$

$$I_o = \frac{2\pi K_3}{m^2 j^3} [u_b N_1(u_b) - u_a N_1(u_a)]K_1 - [u_b J_1(u_b) - u_a J_1(u_a)]K_2 \tag{12.49}$$

The average current density in the tube is the current amplitude divided by the tube wall cross-sectional area

$$J_z(ave) = I_o/S = \frac{I_o}{\pi(a^2 - b^2)} \tag{12.50}$$

12.3   Internal impedance

The internal impedance of the tube is given by (5.75)

$$Z_i(r) = \rho l \frac{J_z(r)}{I_o} \tag{12.51}$$

$$Z_i(r) = \frac{\rho l J_z(u_b)}{I_o} \frac{N_o(u_r)K_1 - J_o(u_r)K_2}{[N_o(u_b)K_1 - J_o(u_b)K_2]} \tag{12.52}$$

$$I_o = \frac{2\pi K_3}{m^2 j^3} [u_b N_1(u_b) - u_a N_1(u_a)]K_1 - [u_b J_1(u_b) - u_a J_1(u_a)]K_2 \tag{12.53}$$

$$K_3 = J_z(u_b)/[N_o(u_b)K_1 - J_o(u_b)K_2] \tag{12.54}$$

$$Z_i(r) = \frac{\rho l m^2 j^3 [N_o(u_r)K_1 - J_o(u_r)K_2]}{2\pi\{[u_b N_1(u_b) - u_a N_1(u_a)]K_1 - [u_b J_1(u_b) - u_a J_1(u_a)]K_2\}}$$

$$\rho l m^2 j^3 = -j\rho l\omega\mu\sigma = -j\omega\mu \ \Omega m^{-1}.$$

$$Z_i(r) = \frac{j\omega\mu[J_o(u_r)K_2 - N_o(u_r)K_1]}{2\pi\{[u_b N_1(u_b) - u_a N_1(u_a)]K_1 - [u_b J_1(u_b) - u_a J_1(u_a)]K_2\}} \ \Omega m^{-1} \quad (12.55)$$

where $K_1 = -[J_1(u_a) - J_1(u_b)]$, $K_2 = -[N_1(u_a) - N_1(u_b)]$ or

$$K_1 = [J_1(u_b) - J_1(u_a)], \quad K_2 = [N_1(u_b) - N_1(u_a)] \quad (12.56)$$

$$Z_i(r) = \frac{j\omega\mu\{[J_o(u_r)[N_1(u_b) - N_1(u_a)] - N_o(u_r)[J_1(u_b) - J_1(u_a)]\}}{2\pi\{[u_b N_1(u_b) - u_a N_1(u_a)]K_1 - [u_b J_1(u_b) - u_a J_1(u_a)]K_2\}} \ \Omega m^{-1} \quad (12.57)$$

$$Z_i(r) = \frac{j\omega\mu}{D}\{[J_o(u_r)[N_1(u_b) - N_1(u_a)] - N_o(u_r)[J_1(u_b) - J_1(u_a)]\} \ \Omega m^{-1} \quad (12.58)$$

where

$$D = 2\pi\{[u_b N_1(u_b) - u_a N_1(u_a)]K_1 - [u_b J_1(u_b) - u_a J_1(u_a)]K_2\} \quad (12.59)$$

12.4   Average current density

The average current density is defined by

$$J_{ave} = \frac{1}{S}\int_s J_z(r)dS \quad (12.60)$$

where S is the cross-sectional area of the current path. Which, in this case, is a tube with cross-sectional area $S = \pi(b^2 - a^2)$. Hence,

$$J_{ave} = \frac{1}{\pi(b^2 - a^2)}\int_a^b J_z(r)2\pi r dr \quad (12.61)$$

From (12.4)

$$u = mr j^{3/2}, \quad r = u/(mj^{3/2}), \quad dr = du/(mj^{3/2}), \quad rdr = udu/(m^2 j^3) \quad (12.62)$$

Hence

$$J_{ave} = \frac{2\pi}{\pi(b^2 - a^2)m^2 j^3}\int_a^b J_z(u)udu \quad (12.63)$$

From (12.77)

$$J_z(u_r) = J_z(u_b)\frac{N_o(u_r)K_1 - J_o(u_r)K_2}{N_o(u_b)K_1 - J_o(u_b)K_2} \quad (12.64)$$

$$J_{ave} = K_4 K_5 [\int_a^b N_o(u_r)K_1 u_r du - \int_a^b J_o(u_r)K_2 u_r du] \quad (12.65)$$

where

$$K4 = J_z(u_b)/[N_o(u_b)K_1 - J_o(u_b)K_2] \quad (12.66)$$

$$K5 = \frac{2\pi}{\pi(b^2 - a^2)m^2 j^3} \quad (12.67)$$

The Bessel function integrals are given by

$$\int uJ_o(u)du = uJ_1(u), \quad \int uN_o(u)du = uN_1(u) \tag{12.68}$$

gives

$$J_{ave} = K_4K_5\left[\int_a^b N_o(u_r)K_1u_r du - \int_a^b J_o(u_r)K_2u_r du\right] \tag{12.69}$$

$$J_{ave} = K_4K_5\{[u_bN_1(u_b) - u_aN_1(u_a)]K_1 - [u_bJ_1(u_b) - u_aJ_1(u_a)]K_2\} \tag{12.70}$$

12.5   Average internal impedance

The average internal impedance of a metal tube length l, supply current amplitude I_o, is given by (5.75)

$$Z_i(ave) = \rho l\frac{J_{ave}}{I_o} \tag{12.71}$$

$$Z_i(ave) = \frac{2\pi\rho lJ_z(u_b)}{\pi(b^2 - a^2)m^2j^3}\frac{\{[u_bN_1(u_b) - u_aN_1(u_a)]K_1 - [u_bJ_1(u_b) - u_aJ_1(u_a)]K_2\}}{I_o[N_o(u_b)K_1 - J_o(u_b)K_2]} \tag{12.72}$$

$$Z_i(ave) = \frac{2\,\rho^2lJ_z(u_b)}{\pi(b^2 - a^2)\omega\mu(-j)}\frac{\{[u_bN_1(u_b) - u_aN_1(u_a)]K_1 - [u_bJ_1(u_b) - u_aJ_1(u_a)]K_2\}}{I_o[N_o(u_b)K_1 - J_o(u_b)K_2]} \tag{12.73}$$

$$Z_i(ave) = \frac{2\rho lJ_z(u_b)}{(u_b^2 - u_a^2)}\frac{\{[u_bN_1(u_b) - u_aN_1(u_a)]K_1 - [u_bJ_1(u_b) - u_aJ_1(u_a)]K_2\}}{I_o[N_o(u_b)K_1 - J_o(u_b)K_2]}, \; \Omega \tag{12.74}$$

since $(b^2 - a^2)m^2j^3 = b^2m^2j^3 - a^2m^2j^3 = u_b^2 - u_a^2$.

$$Z_i(r) = \frac{2\pi\rho l/(m^2j^3)}{[N_o(u_b)K_1 - J_o(u_b)K_2]}$$
$$\times\{[u_bN_1(u_b) - u_aN_1(u_a)]K_1 - [u_bJ_1(u_b) - u_aJ_1(u_a)]K_2\}^{-1}$$

$$Z_i(r) = \frac{\rho l/[N_o(u_b)K_1 - J_o(u_b)K_2]}{\frac{2\pi}{m^2j^3}\{[u_bN_1(u_b) - u_aN_1(u_a)]K_1 - [u_bJ_1(u_b) - u_aJ_1(u_a)]K_2\}}$$

$$Y_i(r) = \frac{\frac{2\pi}{m^2j^3}\{[u_bN_1(u_b) - u_aN_1(u_a)]K_1 - [u_bJ_1(u_b) - u_aJ_1(u_a)]K_2\}}{\rho l/[N_o(u_b)K_1 - J_o(u_b)K_2]} \tag{12.75}$$

$$J_{zr} = AJ_o(u) + BN_o(u) \tag{12.76}$$

$$J_z(u_r) = J_z(u_b)\frac{N_o(u_r)K_1 - J_o(u_r)K_2}{N_o(u_b)K_1 - J_o(u_b)K_2} \tag{12.77}$$

$$B_{zr} = B_{zb}\frac{N_rK_1 - J_rK_2}{N_bK_1 - J_bK_2} \tag{12.78}$$

where $K_1 = J_a' - J_b'$, $K_2 = N_a' - N_b'$ are constants.

Substituting (13.37) and (12.78) into (12.31) gives

$$Z_\phi = \rho m j^{3/2} \frac{K_1 N_r' - K_2 J_r'}{K_1 N_r - K_2 J_r}$$

(12.79)

12.6   The average magnetic field and permeability

Previously for a *solid* conducting cylinder, the average permeability was defined by (see section 8.3)

$$\mu = \frac{\bar{B}}{B_a} = \frac{2}{a^2 B_a} \int_0^a r B(r) dr$$

(12.80)

where at the cylinder surface $B_a = B(a) = \mu_o H_a$.

For a conducting tube the average magnetic field is

$$\bar{B} = \frac{2}{b^2 - a^2} \int_a^b r B(r) dr$$

(12.81)

If the magnetic field in the tube walls B_i is uniform and constant then the average internal field is as expected

$$\bar{B} = \frac{2}{b^2 - a^2} B_i \int_a^b r dr = \frac{2}{b^2 - a^2} B_i \frac{b^2 - a^2}{2} = B_i$$

(12.82)

The average permeability of the tube is then

$$\bar{\mu} = \frac{\bar{B}}{B_b} = \frac{2}{B_b(b^2 - a^2)} \int_a^b r B(r) dr$$

(12.83)

where B_b is the applied field. Substitute for $B_r = B(r)$ (12.78)

$$\bar{\mu} = \frac{2}{(b^2 - a^2)} \int_a^b r \frac{N_r K_1 - J_r K_2}{N_b K_1 - J_b K_2} dr$$

(12.84)

12.7   Field transmission and screening factor

The *Field Transmission* (T) is defined by the ratio

$$T = \frac{B(b)}{B(a)}$$

(12.85)

Also of interest is the *screening factor*

$$S = 1 - T = 1 - \frac{B(b)}{B(a)}$$

(12.86)

12.8   Admittance analysis

The internal admittance of a solid cylindrical conductor of radius a is Y_a. The internal admittance of a solid cylindrical conductor radius of b is Y_b. The internal admittance of a tube thickness $a - b$ is $Y_{ab} = Y_a - Y_b$.

Now for a solid cylindrical conductor, the radius a the internal impedance is

$$Z_a = R_{dc} \frac{u_a}{2} \frac{J_o(u_a)}{J_1(u_a)} \tag{12.87}$$

where $u_a = j^{3/2} 2^{1/2} a/\delta$ and $\delta = (2\rho/(\omega\mu))^{1/2}$ is the skin depth. Similarly for a solid cylindrical conductor radius b the internal impedance is

$$Z_b = R_{dc} \frac{u_b}{2} \frac{J_o(u_b)}{J_1(u_b)} \tag{12.88}$$

where R_{dc} is the steady state resistance of the conductor

$$R_{dc} = \frac{\rho l}{\pi r^2} \tag{12.89}$$

where ρ is the resistivity of the conductor and l its length.

The admittance is the reciprocal of these impedances. Hence,

$$Y_{ab} = 1/Z_{ab} = \frac{2\pi}{\rho l} \left[\frac{a^2 J_1(u_a)}{u_a J_o(u_a)} - \frac{b^2 J_1(u_b)}{u_b J_o(u_b)} \right] \tag{12.90}$$

At high frequencies

$$\frac{J_1(u_r)}{J_o(u_r)} = -1/j \tag{12.91}$$

Hence,

$$Y_{ab} = -\frac{2\pi}{j\rho l} \left[\frac{a^2}{u_a} - \frac{b^2}{u_b} \right] \tag{12.92}$$

and, we can put

$$Y_{ab} = K_f / f^{1/2} \tag{12.93}$$

$$Z_{ab} = 1/Y_{ab} = f^{1/2} / K_f \tag{12.94}$$

where

$$K_f = \frac{a - b}{l} \sqrt{\frac{2\pi}{j\rho\mu}} \tag{12.95}$$

Chapter 13
Hollow cylindrical conductor – axial B

13.1 Introduction

A magnetic field applied to a *solid* cylindrical conductor was considered previously, page 131. In this chapter, we consider a conducting tube such as may be used in sub-station bus conductors [109](not a clippie!). The present analysis and measurements are mainly limited to low-power conditions.

13.2 Magnetic field analysis

In this case, we consider a time-dependent magnetic field $\mathbf{B} = \mathbf{B}_o e^{j\omega t}$ applied axially to a long cylindrical tube conductor with inner radius a, and outer radius b (Figure 13.1). $\mathbf{B}_o$ is less than the saturation flux density $\mathbf{B}_{sat}$ and any static magnetic field is assumed zero, that is, $\mathbf{B}_{dc} = 0$. The field penetration into the conductor is obtained from Maxwell's equations, which yield the diffusion (5.3) re-written here as follows:

$$\nabla^2 \mathbf{B} = jm^2 \mathbf{B} \tag{13.1}$$

where $m^2 = \omega\mu\sigma$ or $m = \sqrt{2}/\delta$, skin depth $\delta = \sqrt{(2/\omega\mu\sigma)}$. In the following, the magnetic field is in the z-direction, so $\mathbf{B} = B\mathbf{a}_z$, $\rho = r$, and we drop the subscript z. In cylindrical co-ordinates (13.1), then becomes

$$\nabla^2 B = r^{-1}\frac{\partial}{\partial r}\left(r\frac{\partial B}{\partial r}\right) + r^{-2}\frac{\partial^2 B}{\partial \phi^2} + \frac{\partial^2 B}{\partial z^2} = jm^2 B. \tag{13.2}$$

Assuming B does not vary with angle ϕ or distance z, then

$$r^2\frac{\partial^2 B}{\partial r^2} + r\frac{\partial B}{\partial r} - jm^2 B r^2 = 0 \tag{13.3}$$

which is Bessel's modified equation order 0. The general solution for the magnetic field inside the tube walls is

$$B(r) = AJ_o(u_r) + BN_o(u_r) \tag{13.4}$$

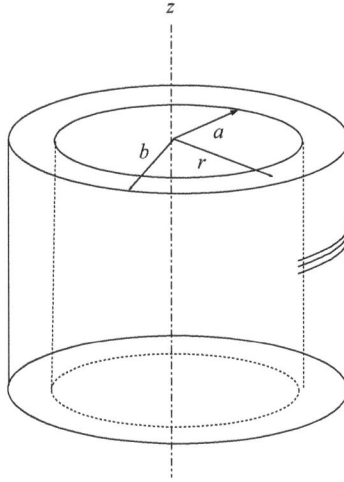

Figure 13.1   Magnetic field applied in the z-axial direction of a metal tube

where $a \leq r \leq b$, $J_o(u)$, and $N_o(u)$ are Bessel's functions of the first kind and second kind respectively with complex argument u and order zero. (See the general Appendix.) For the outer surface of the cylinder $r = b$, the solution is

$$B(b) = AJ_o(u_b) + BN_o(u_b) \tag{13.5}$$

For the inner surface of the cylinder $r = a$, the solution is

$$B(a) = AJ_o(u_a) + BN_o(u_a) \tag{13.6}$$

where both $B(a)$ and $B(b)$ are constants for a given frequency. Solving (13.5) and (13.6) for A and B

$$A = [B(b)N_o(u_a) - B(a)N_o(u_b)]/D \tag{13.7}$$

$$B = [B(a)J_o(u_b) - B(b)J_o(u_a)]/D \tag{13.8}$$

$$D = J_o(u_b)N_o(u_a) - J_o(u_a)N_o(u_b) \tag{13.9}$$

Although we know $B(b)$ because this is the applied field. $B(a)$ is unknown but can be obtained by differentiating $B(b)$ and $B(a)$ wrt u_b and u_a as indicated by the prime (′).

$$B'(b) = 0 = AJ'_o(u_b) + BN'_o(u_b) \tag{13.10}$$

$$B'(a) = 0 = AJ'_o(u_a) + BN'_o(u_a) \tag{13.11}$$

both $B'(a)$ and $B'(b)$ are zero because these fields are constants at the surfaces a and b. Hence,

$$AJ'_o(u_b) + BN'_o(u_b) = AJ'_o(u_a) + BN'_o(u_a) \tag{13.12}$$

$$\frac{A}{B} = \frac{N'_o(u_a) - N'_o(u_b)}{J'_o(u_b) - J'_o(u_a)} \tag{13.13}$$

For the general solution (13.4) and (13.5) substitute for A and B from (13.7) and (13.8), and abbreviating

$J_o(u_r) = J_r$, $J_o(u_a) = J_a$, $J_o(u_b) = J_b$
$N_o(u_r) = N_r$, $N_o(u_a) = N_a$, $N_o(u_b) = N_b$
$B(b) = B_{zb}$, $B(a) = B_a$, $B(r) = B_{zr}$

$$B_{zr} = \frac{B_{zb}N_a - B_{za}N_b}{D}J_r + \frac{B_{za}J_b - B_{zb}J_a}{D}N_r \tag{13.14}$$

$$B_{zb} = \frac{B_{zb}N_a - B_{za}N_b}{D}J_b + \frac{B_{za}J_b - B_{zb}J_a}{D}N_b \tag{13.15}$$

where

$$D = J_bN_a - J_aN_b \tag{13.16}$$

From (13.14) and (13.15), put

$$c_1 = a_1x + b_1y \tag{13.17}$$

$$c_2 = a_2x + b_2y \tag{13.18}$$

where

$c_1 = B_{zb}, a_1 = J_b, x = A, b_1 = N_b, y = B,$
$c_2 = B_{zr}, a_2 = J_r, x = A, b_2 = N_r, y = B$ $\tag{13.19}$

Solving (13.17) and (13.18) for x and y

$$x = \frac{c_1b_2 - c_2b_1}{a_1b_2 - a_2b_1} \tag{13.20}$$

$$y = \frac{a_1c_2 - a_2c_1}{a_1b_2 - a_2b_1} \tag{13.21}$$

Substituting for the parameters from (13.19) gives

$$A = \frac{B_{zb}N_r - B_{zr}N_b}{J_bN_r - J_rN_b} \tag{13.22}$$

$$B = \frac{B_{zr}J_b - B_{zb}J_r}{J_bN_r - J_rN_b} \tag{13.23}$$

Hence,

$$\frac{A}{B} = \frac{B_{zb}N_r - B_{zr}N_b}{B_{zr}J_b - B_{zb}J_r} = \frac{N'_a - N'_b}{J'_b - J'_a} \tag{13.24}$$

Solving for B_{zr} gives

$$B_{zr} = B_{zb}\frac{N_rK_1 - J_rK_2}{N_bK_1 - J_bK_2} \tag{13.25}$$

Obtaining the surface magnetic field at a from (12.27), where N_r becomes N_a and J_r becomes J_a gives

$$B_{za} = B_{zb}\frac{N_aK_1 - J_aK_2}{N_bK_1 - J_bK_2} \tag{13.26}$$

where $K_1 = J'_a - J'_b$, $K_2 = N'_a - N'_b$ are constants for fixed frequency.

13.3   Current density

For a cylindrical conductor in an axial magnetic field the current density is (see Section 8.3.4)

$$J_\phi = \frac{1}{\mu_o}\frac{\partial B(r)}{\partial r} \tag{13.27}$$

For a conducting tube

$$B(u) = AJ_o(u) + BN_o(u) \tag{13.28}$$

where

$$u = mrj^{3/2}, m = \sqrt{\omega\sigma\mu}, \frac{du}{dr} = mj^{3/2} \tag{13.29}$$

$$J_\phi = \frac{1}{\mu_o}\frac{\partial B(r)}{\partial r} = \frac{1}{\mu_o}\frac{\partial B(u)}{\partial u}\frac{du}{dr} \tag{13.30}$$

$$J_\phi = \frac{mj^{3/2}}{\mu_o}\frac{\partial B(u)}{\partial u} \tag{13.31}$$

From (13.25), we can put

$$\frac{\partial B(u)}{\partial u} = B_b\frac{\partial}{\partial u}\left[\frac{N_rK_1 - J_rK_2}{N_bK_1 - J_bK_2}\right] = B_b\frac{\partial F(u)}{\partial u} \tag{13.32}$$

where

$$F(u) = \left[\frac{N_rK_1 - J_rK_2}{N_bK_1 - J_bK_2}\right] \tag{13.33}$$

Now for a given frequency, J_a', J_b', N_a', N_b', are all constants (see abbreviations page 167). Hence, we can put

$$\frac{\partial F(u)}{\partial u} = \frac{\partial}{\partial u}\frac{K_1N_r - K_2J_r}{N_bK_1 - J_bK_2} \tag{13.34}$$

Hence (13.32) becomes

$$\frac{\partial B(u)}{\partial u} = B_b\frac{\partial}{\partial u}\frac{K_1N_r - K_2J_r}{N_bK_1 - J_bK_2} \tag{13.35}$$

where $K_1 = J_a' - J_b'$, $K_2 = N_a' - N_b'$ are constants.
Equation (13.31) becomes

$$J_\phi = \frac{B_b mj^{3/2}}{\mu_o}\frac{\partial}{\partial u}\left[\frac{K_1N_r - K_2J_r}{N_bK_1 - J_bK_2}\right] \tag{13.36}$$

or

$$J_\phi = \frac{B_b mj^{3/2}}{\mu_o}\left[\frac{K_1N_r' - K_2J_r'}{N_bK_1 - J_bK_2}\right] \tag{13.37}$$

where the primes (') refer to the derivative wrt u.

$$J_\phi = \frac{B_b mj^{3/2}}{\mu_o(N_bK_1 - J_bK_2)}\frac{\partial(K_1N_r - K_2J_r)}{\partial u} \tag{13.38}$$

$$J_o(u_r) = J_r, J_o(u_a) = J_a, J_o(u_b) = J_b,$$
$$N_o(u_r) = N_r, N_o(u_a) = N_a, N_o(u_b) = N_b,$$
$$B(b) = B_b, B(a) = B_a, B(r) = B_r$$

$$B'(u) = AJ_o'(u) + BN_o'(u) \tag{13.39}$$

Using the identity

$$\frac{dJ_o(u)}{du} = -J_1(u) \tag{13.40}$$

and noting that $-j^{3/2} = j^{1/2}$

$$J_\phi = \frac{mj^{1/2}}{\mu_o}[AJ_1(u) + BN_1(u)] \tag{13.41}$$

13.3.1   Impedance

The wave impedance is defined by

$$Z_\phi = \frac{E_\phi}{H_z} \tag{13.42}$$

where E_ϕ is given by

$$E_\phi = \rho J_\phi \tag{13.43}$$

ρ is the resistivity of the metal tube. J_ϕ is given by (13.37) and H_z by (13.25).

$$B_r = B_b F(u) = B_b \frac{K_1 N_r - K_2 J_r}{N_b K_1 - J_b K_2} \tag{13.44}$$

where $K_1 = J_a' - J_b', K_2 = N_a' - N_b'$ are constants.
Substituting (13.37) and (13.44) into (13.42) gives

$$Z_\phi = \rho m j^{3/2} \frac{K_1 N_r' - K_2 J_r'}{K_1 N_r - K_2 J_r} \tag{13.45}$$

13.4   The average magnetic field and permeability

Previously for a *solid* conducting cylinder, the average permeability was defined by (see Section 8.3)

$$\mu = \frac{\bar{B}}{B_a} = \frac{2}{a^2 B_a} \int_0^a rB(r)dr \tag{13.46}$$

where at the cylinder surface $B_a = B(a) = \mu_o H_a$.
For a conducting tube, the average magnetic field is

$$\bar{B} = \frac{2}{b^2 - a^2} \int_a^b rB(r)dr \tag{13.47}$$

To prove this, if the magnetic field in the tube walls B_i is uniform and constant, then the average internal field is

$$\bar{B} = \frac{2}{b^2 - a^2} B_i \int_a^b r\,dr = \frac{2}{b^2 - a^2} B_i \frac{b^2 - a^2}{2} = B_i \tag{13.48}$$

as expected. The average permeability of the tube is then

$$\bar{\mu} = \frac{\bar{B}}{B_b} = \frac{2}{B_b(b^2 - a^2)} \int_a^b rB(r)\,dr \tag{13.49}$$

where B_b is the applied field. Substitute for $B_r = B(r)$ from (13.44)

$$\bar{\mu} = \frac{2}{(b^2 - a^2)} \int_a^b r \frac{N_r K_1 - J_r K_2}{N_b K_1 - J_b K_2}\,dr \tag{13.50}$$

13.5   Azimuthal impedance

Similar analysis applied to a conducting tube, outside radius a and inside radius b, gives for the azimuthal impedance ·

$$Z = V/I = \frac{2\pi\rho}{b} \frac{[u_a J_1(u_a) - u_b J_1(u_b)]}{[J_o(u_a) - J_o(u_b)]} \tag{13.51}$$

For $b = 0$ this reduces to (9.1), the case of a solid cylinder.

13.6   Power dissipation

The time averaged power dissipation is (see page 59, (3.102)),

$$<P> = \frac{I_o^2}{2} Re(Z) = \frac{V_o^2}{2} Re(1/Z),\ W \tag{13.52}$$

For the power dissipation in a hollow tube, substituting for the impedance (13.51) gives

$$<P> = H_a^2 b\pi\rho \left| \frac{J_o(u_a) - J_o(u_b)}{J_o(u_a)} \right|^2 Re\left[\frac{u_b J_1(u_b) - u_a J_1(u_a)}{J_o(u_a) - J_o(u_b)} \right],\ W \tag{13.53}$$

where H_a is the applied field in A/m.

Chapter 14

Magnetic field penetration into a copper tube: experimental measurements

14.1 Introduction

This section presents experimental measurements of magnetic field penetration through the walls of a conducting cylinder. The field was produced by a solenoidal wire coil wound on a plastic tube, which could be slid over a copper tube, Figure 14.1. These experiments were carried out with the object of comparing experimental measurements with theoretical calculations for the case of a sinusoidal magnetic field applied parallel to the axis of a conducting tube. The field was produced by a solenoidal wire coil driven from a Function Generator HP3325A, a power amplifier Sherwood AX4050R, and an 11 Ω, 50 W series resistor (two parallel 22 Ω ±5%, 25 W resistors) as shown in Figures 14.2, 14.3, and 14.4.

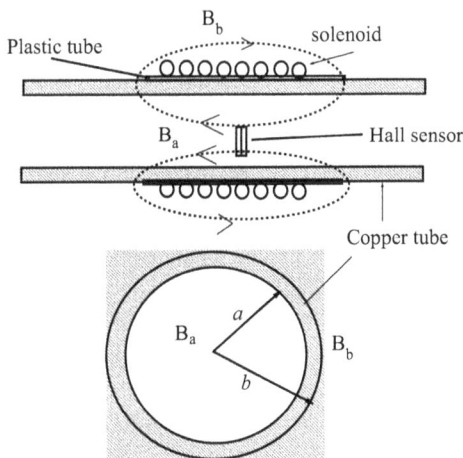

Figure 14.1   The Hall sensor was used to measure the solenoid magnetic fields either outside (B_b) or inside (B_a) the copper or plastic tubes

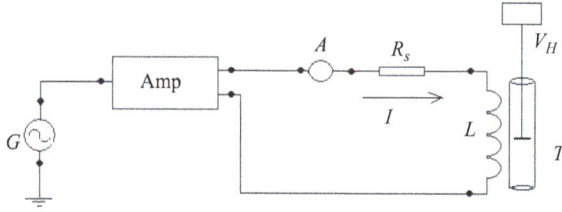

*Figure 14.2   Circuit for measuring the field penetration in a metal tube T.
             G-function generator HP3325A, Amp-Power amplifier Sherwood
             AX4050R, A-Solartron DMM 7150+, Rs = 11 Ω, 50 W, L = solenoidal
             field coil, V_H – Hall sensor and Solartron DMM 7150.*

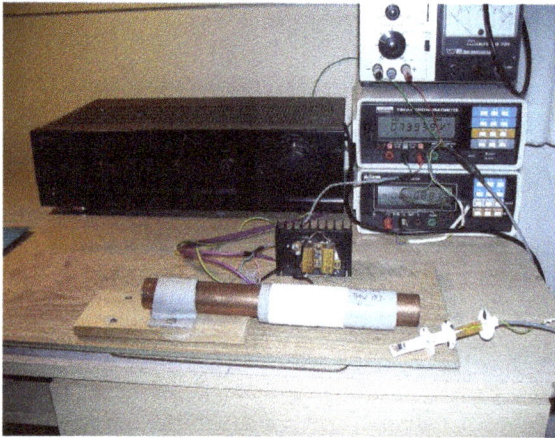

*Figure 14.3   Experimental arrangement for the measurement of coil two magnetic
             field. Shows the Sherwood Stereo amplifier, power supply, series
             resistor and copper tube, and plastic tube supporting the solenoid.
             The Hall sensor is shown on the bottom rhs. This could be slid into
             either the copper or plastic tubes.*

*Figure 14.4   Photograph showing dual differential Hall sensor and axial inductor
             sensor mounted on PCB strip. The PTFE covered plastic discs
             allowed the probe to be positioned into the centre of the coil or the
             copper tube.*

14.2   Magnetic field coil details

The flux density at the centre of a solenoid of finite length is [111] p233 Kraus

$$B = \frac{\mu_o N I}{\sqrt{4R_c^2 + L_c^2}} \tag{14.1}$$

where I is the current in amps, $\mu = \mu_o \mu_r$, where $\mu_r =$ relative permeability of the tube = 1 in this case, N = number of turns, R_c is the coil radius, $L_c =$ coil length. The coil inductance is

$$L = NBA/I \tag{14.2}$$

In the experiments, the coil (designated C2) consisted of a 0.3 mm diameter copper wire wound on a grey plastic tube Tp and fixed with PTFE tape. This could be slid over a copper tube T4 to produce the required axial field. The number of turns $N = 153$, length $L_c = 60$ mm, diameter $D_c = 34.6$ mm. The copper tube, designated here T4, had length $l_T = 250$ mm, diameter $d_T = 28$ mm, and wall thickness $t_w = 1$ mm.

    The inductance and resistance of the coil wound only on the plastic tube was measured with an Inductance capacitance resistance (LCR) meter Racal-Dana Databridge 9343M. The results are given in Table 14.1.

    Substituting for (14.1), coil turns and dimensions gives for $I = 1A$ $B = 2.776$ mT, $L = 399.3$ μH. This dc inductance agrees well with the low frequency measured values given in Table 14.1. The coil was also measured with a bridge WK B424 at 1 kHz. This gave $L = 393$ μH, $R = 4.16$ Ohms which also agrees approximately with the databridge measurement at 1 kHz.

14.2.1   Coil temperature

During extended periods of measurement, the temperature of the coil increased from room temperature, 20–40°C. The temperature was measured about halfway along the coil with a thermocouple trapped between the coil's plastic former and the copper tube.

14.2.2   DC Test of Coil 2 and Hall sensor

The magnetic field was measured using a Hall-effect IC sensor, UGN-3503U, and a digital multimeter (DMM) Solartron 7150 plus. Results were obtained using a double differential arrangement with two back-to-back sensors. The coil 2 was connected

Table 14.1   *L and R of coil on plastic tube Tp using LCR meter Racal-Dana Databridge 9343M.*

f (Hz)	*L* (μH)	*R* (Ω)	Reverse *L* (μH)	Reverse *R* (Ω)	Ave *L* (μH)	Ave *R* (Ω)
100	400	4.161	401	4.165	400	4.162
1k	396	4.188	396	4.190	396	4.189
10k	388	4.466	388	4.457	388	4.462

Table 14.2   DC Current through Coil 2 and measured Hall voltage using differential Hall probe

I (A)	0	0.1	0.2	0.3	0.4	0.5	0.6	0.7	0.8	0.9	1
V_H (mV)	46.2	55.3	64.6	73.7	82.9	92.2	101.2	110.3	119.4	128.1	137.6

directly to a Farnell Power Supply XA35-2T with integral current and voltage meters. The coil was empty except for the Hall sensor placed midway inside the coil. The results are shown in Table 14.2 [112].

To obtain the magnetic field B2 (14.1) was used. This gives

$$B_2 = 2.776I \, \text{mT}. \tag{14.3}$$

The Hall voltage curve fit equation from Table 14.2 and Figure 14.5 is

$$V_H = 91.31I + 46.3 \, \text{mV}. \tag{14.4}$$

Substituting for I from (14.3) gives

$$V_H = 32.96B_2 + 46.3 \, \text{mV} B_2 = 0.0304V_H - 1.4076 \, \text{mT} \tag{14.5}$$

For example, if $I = 0.6$ A, $V_H = 100$ mV, then $B_2 = 1.63$ mT agreeing with Figure 14.5. This gives the sensitivity of the Hall measurement to be 32.96 mV/mT and the magnetic field sensitivity as 0.0304 mT/mV.

For a single Hall device used, UGN-3503U, the data sheet gave the sensor sensitivity dV as minimum = 0.75, typical = 13, and maximum = 17.2 mV/mT. We used two sensors in a differential connection, which yielded $2dV = 32.96$ mV/mT, which is close to the manufacture's max data of 34.4 mV/mT. Note that in Figure 14.5 there is a relatively large Hall offset voltage $V_H(0) = 46.36$ mV for zero current. This is absent in AC measurements.

14.2.3   AC test of coil 2 and Hall sensor: no copper tube

In this case, coil 2 was driven from a sinusoidal source, as shown in Figure 14.2. This consisted of a function generator FG (HP3325A) connected to the 'Right' phono input of a stereo power amplifier PA (Sherwood AX4050). The 'Right' output of this was connected to a series resistance of 11 Ohms, 50W. This consisted of two 22 Ohms 5%, 25W power resistors connected in parallel and mounted on a heat sink, as shown in figure 14.3.

Coil 2 was connected to one side of the power resistor and the PA ground. The coil current and Hall voltages were measured with Solartron DMMs, Sol +, and Sol, respectively. The magnetic field was measured midway in coil 2 with no copper tube, using a purpose-made differential Hall sensor with 6 V P/S and an inductor sensor adjacent to the Hall sensor, Figure 14.4.

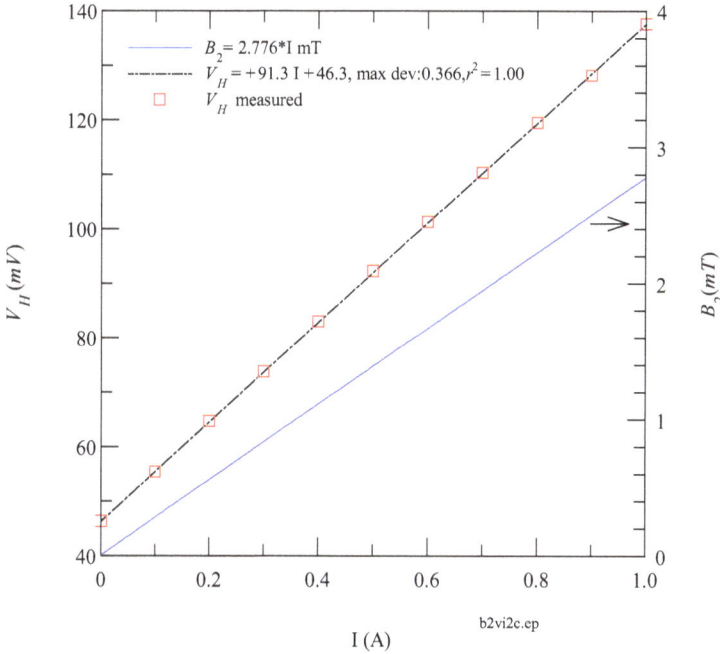

Figure 14.5 *DC Hall sensor voltage V_H and calculated magnetic field B2 as a function of coil 2 current. Using MSR Hall probe and 6 V power supply, B2 from (14.3). All currents and voltages are dc.*

Table 14.3 *Axial lead inductor, Painton 69.47, 1 mH. L and R measurements using databridge LCR meter*

f (Hz)	L (μH)	$R\Omega$	Revers L (μH)	Revers $R\Omega$	Ave L (uH)	Ave $R\Omega$
100	1050	55.71	1040	55.73	1045	55.72
1k	1037	55.74	1036	55.74	1036	55.74
10k	1033	55.86	1032	56.0	1032	55.93

14.2.4 Inductor magnetic field sensor

This is a commercial axial lead inductor, Painton 69.47, 1 mH length 8.5 mm, and 3.22 mm diameter. This includes magnetic material, presumably ferrite. The low-frequency L and R were measured using the databridge LCR meter and the results listed in Table 14.3.

The results obtained are shown in Figure 14.6. These were obtained with an FG output of 2 mV, frequency $f = 1$ kHz supplying the right phono input of the PA. The output current of the PA was adjusted with the PA volume control, and the current was monitoring with the DMM.

b2vh2b.ep

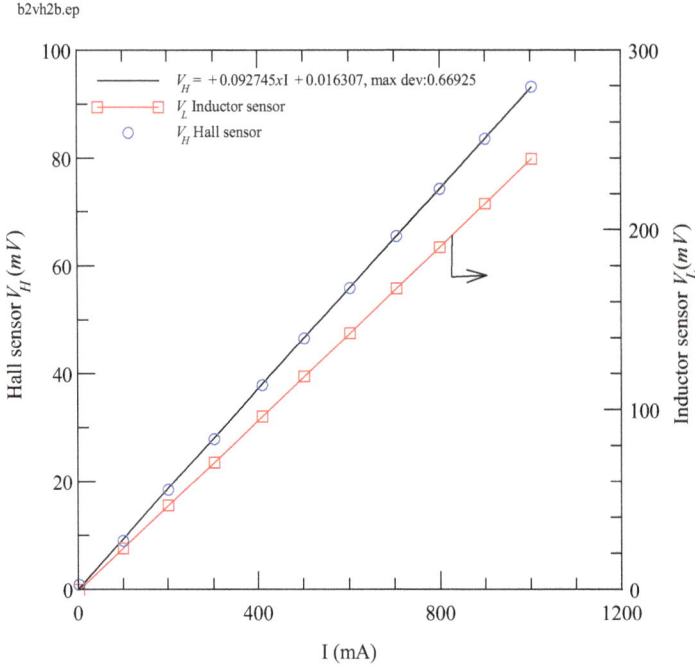

Figure 14.6   AC measurements of the magnetic field in the centre of coil 2 using a differential Hall sensor and Inductor sensor

The Hall voltage curve fit, equation from Figure 14.6, is

$$V_H = 92.7I + 0.0163 \,\text{mV},\tag{14.6}$$

where I is in Amps and V_H in mV and both are rms values. Assuming that at this frequency the field is still given by (14.3) (since the inductance measured at 1 kHz is close to the calculated value using $B2$), then

$$V_H = 33.39B_2 + 0.0163 \,\text{mV} \quad B_2 = 0.03V_H \,\text{mT}\tag{14.7}$$

This gives the sensitivity of V_H to B as 33.39 mV/mT and B to V_H as 0.03 mT/mV. Figure 14.7 shows the good agreement obtained for $B2$ determined from V_H measurements and $B2$ calculated from the current in the coil.

14.2.5   Axial sinusoidal B field applied to copper tube T4

These measurements used the same experimental arrangement as that described previously. In the present experiment, coil 2 was placed over the copper tube T4, and the frequency was varied for constant coil current.

The results obtained are shown in Figures 14.8 and 14.9. These were obtained with an FG output of 2 mV supplying the right phono input of the PA. The output current of the PA was kept constant to within less than 1% at each frequency by manually adjusting the PA volume control and monitoring the current in the DMM.

Magnetic field calculated from RMS current and measured V_H for coil 2

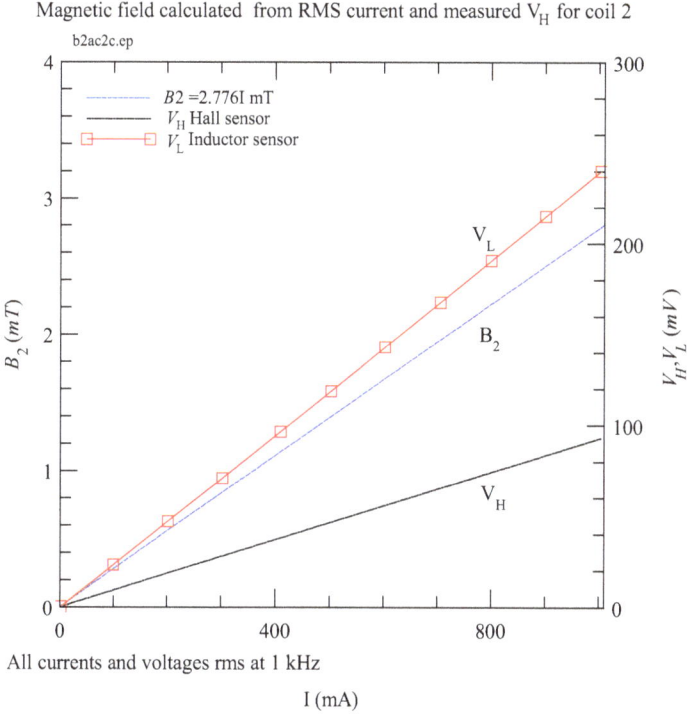

All currents and voltages rms at 1 kHz

I (mA)

Figure 14.7   Comparison of ac magnetic field measured at 1 kHz using the Hall sensor and magnetic field calculated from the current in the coil. Also shown is the inductor sensor voltage VL.

14.3   Field transmission and screening factor

The field transmission coefficient (T) is defined by the ratio field inside tube over Field outside tube. Thus,

$$T = \frac{B(a)}{B(b)} \tag{14.8}$$

where $B(a)$ is the field at the tube inner surface $r = a$ and $B(b)$ is the field at the tube outer surface $r = b$.

Also of interest is the *screening factor*

$$S = 1 - T = 1 - \frac{B(a)}{B(b)} \tag{14.9}$$

The fraction of the field penetration between the inner surface of the tube and a point radius r in the tube is $T_r = B_r/B_b$.

The transmission coefficient T and screening factor S for Cu tube T4 are shown in Figure 14.10. These were obtained from Figure 14.8 and (13.26), p. 185.

Hall field measurement and Transmission coefficient T for Cu tube T4

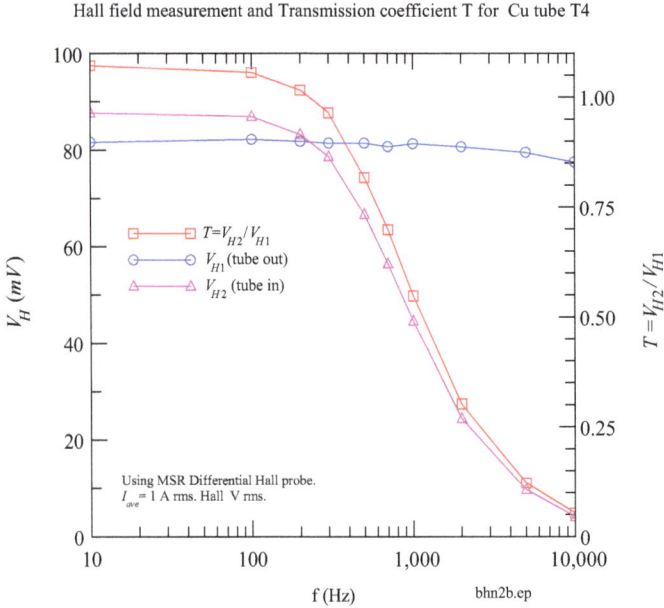

Figure 14.8   Hall voltage as a function of frequency with the Hall sensor positioned mid-way in Coil 2, with and without the copper tube T4. The current was maintained constant at 1 A within less than 1% for each frequency. Also shows the transmission coefficient T.

14.4   Field errors

The experimental result for the transmission coefficient T differs from the theoretical result as shown in Figure 14.10. In these measurements the tube length was much larger than the solenoid length. Hence, the field at the end of the solenoid had a vertical component. The relationship between the horizontal susceptibility (h) component and vertical susceptibility component (v) is [12]

$$\chi_h = \frac{1}{2}\chi_v \tag{14.10}$$

$$\mu_h - 1 = \frac{1}{2}(\mu_v - 1) \tag{14.11}$$

The mean permeability is (13.46)

$$\mu_{av} = \frac{B_{av}}{B_b} \tag{14.12}$$

Hence,

$$\frac{B_v}{B_b} = 2\frac{B_h}{B_b} - 1 \tag{14.13}$$

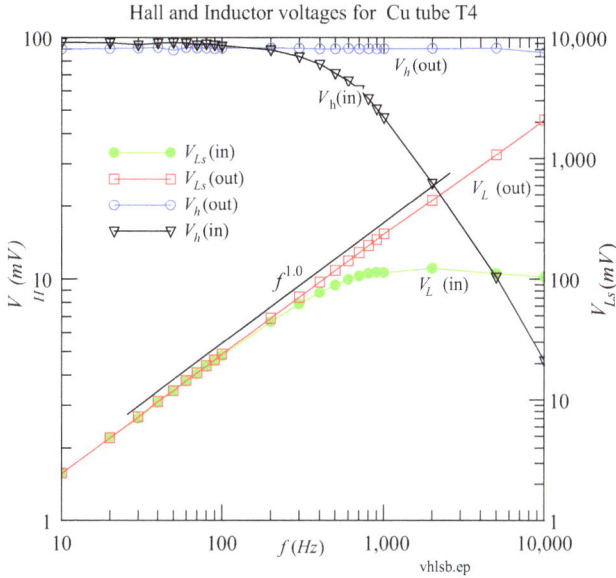

Figure 14.9   Hall voltage V_H and inductor sensor voltage V_{Ls} with copper tube present (in) and copper tube absent (out)

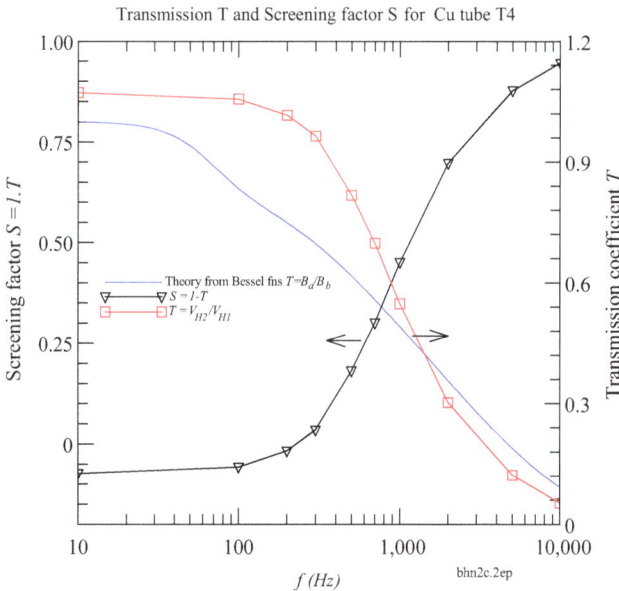

Figure 14.10   Transmission coefficient T and screening factor S for Cu tube T4. From Figure 14.8 and the theoretical T from (13.26), p. 185

The Hall device was positioned in the centre of the coil close to the surface, so it detected the field parallel to the surface. This field was constant and uniform across the diameter of the coil. The vertical field at the solenoid end was also probed using the Hall sensor and found to be only a few percent of the parallel field inside the solenoid. The theoretical result depends on Bessel functions and their derivative (see (13.26), p. 185, and the MATLAB® program below). Also, the tube radii a and b are close to each other, so any small differences may cause large errors in the theoretical results.

14.5   MATLAB® program for transmission coefficient, T

Note: The! symbol substitutes for the percentage symbol used for comments in MATLAB®.

```
!Btube.txt Matlab analysis of tube flux penetration. format short e !for x = linspace(.0001,.004,50) for x = logspace(1,4,100) f=x; w=2*pi*f; muo=pi*4e-7; mur=1; mu=muo*mur; rho=1.7e-8; !ohm m. b=14e-3; !tube od radius, m. a=13e-3; !tube id radius, m.

Joa=besselj(0,ua); Job=besselj(0,ub);

J1a=besselj(1,ua); J1b=besselj(1,ub);

Noa=bessely(0,ua); Nob=bessely(0,ub);

N1a=bessely(1,ua); N1b=bessely(1,ub);

dJ1=J1b-J1a; dN1=N1b-N1a;

BaN=(Noa*dJ1+Joa*dN1/(Nob*dJ1+Job*dN1); !Ba inside hollow tube BaN=Ba/Bb
BN=Joa/Job; !Solid tube BN= Ba/Bb !See equation 13.26, p185. muave=(2/ub)*J1b/Job;
!ave perm mu=Bave/Bb

disp([x abs(BaN) abs(BN)]) !disp([x real(B1N) imag(B1N) real(BN) imag(BN)])
!disp([x abs(muave)])

end
```

Chapter 15
Impedance measurement techniques

15.1 Introduction

In skin effect measurements, the impedance may be very low particularly at frequencies in the audio and power range, where the ac resistance can be less than a milliohm, and accurate measurements become difficult to achieve [119]. Hence, in order to validate the skin effect theory based on impedance measurements, sensitive instruments are required.

Some of the earliest impedance measuring techniques used the principle of the balanced bridge as used for resistance measurements. This also included the transformer ratio arm bridges as used by Wayne Kerr in the 'Universal Bridge' B224 (200 Hz to 50 kHz) and the RF Bridge B601 (15 kHz to 5 MHz) [116]. The bridge techniques require laborious balancing but can be very accurate and are commonly used in the laboratory for precision measurements.

For direct variable frequency measurements of L, C, and R, techniques are required that measure the vector current through the device and vector voltage across it, $\mathbf{Z} = \mathbf{V}/\mathbf{I}$. This is referred to as the I, V method of measuring the complex impedance of a device from which L, C, and R may be determined. For high-frequency measurements above about 1 MHz, the reflection and /or transmission coefficient parameter values may be measured and related to the impedance of the device under test as in the HP 4191A and Agilent 4395A Network Analysers. Further details about these techniques may be obtained from the Impedance Measurement Handbook [117,118].

15.2 Recent developments

During the past few years, there has been an increase in the development of novel impedance measuring techniques. This has been largely stimulated by clinical applications in the fields of bio-impedance and impedance tomography. This includes a high speed bio-impedance spectrometer based on a Field Programmable Gate Array (FPGA) [121], fast impedance measurements using curve fitting algorithms [122], impedance measurements using a digital signal processor (DSP) [123], detection of magnetic fields attenuated by the skin effect using a DSP [124], a high speed impedance measuring system based on information filtering demodulation

[125], Complex impedance measuring system based on the $I - -V$ method [126], genetic algorithm method [127], impedance spectroscopy using broadband excitation [128,129], impedance measurements using a gain phase meter (GPM), [104] and impedance measurements using the three voltmeter method (3VM), [130,131]. Although at present the GPM technique relies on bench instruments, further developments are possible using the single chip RF/IF Gain and Phase detector system AD8302 [133], which should permit low-cost measurements of impedance in the field for frequencies up to 2.5 GHz. The introduction of a high-precision impedance converter chip, which includes a programmable tunable frequency generator with a 12- bit, 1 MSPS (AD5933) or 250 kSPS (AD5934) analog-to-digital converter is also leading to impedance measurement field applications [132].

15.3   GPM technique

In addition to the other techniques described above, a GPM technique was also employed here to measure impedance as a function of frequency. This was particularly useful for the low-frequency skin effect, where the resistance may be less than a milliohm. Novel software routines were produced to extract the complex components of the impedance from the signal amplitude and phase. The technique is fully described in Raven [120]. The paper initially presents an analysis of the technique, followed by details of the de-embedding procedure. A separate alternative circuit and analysis are described for the measurement of capacitive impedance. The results of the experimental measurements are divided into two parts: firstly, measurements over a continuous frequency range 100 Hz to about 10 MHz using reference samples, which demonstrate the accuracy of the technique. The reference samples included R, L, and C components and reverse-biased p–n junctions. The latter was obtained at 1 MHz with an amplitude of 15.5 mV rms, sufficiently low to only slightly modulate the depletion region. Finally, results were presented that show that the technique is sufficiently sensitive to measure the skin effect in a short copper rod at audio frequencies to about 10 MHz. These results are found to compare well with theoretical analysis using Bessel functions, including the possible detection of the internal inductance of the rod, which is significantly less than the external inductance.

15.3.1   Measurement system and procedures

The test system used a HP3575A GPM, Fig. 15.1. This includes two independent input channels A and B with sensitivity of 0.2 mV to 20 V rms in two ranges, frequency response of 1 Hz to 13 MHz in four overlapping ranges. The input impedance of each channel is 1 Megohm in parallel with less than 30 pF. The results reported here were obtained with the instrument set to 0.2 mV to 2 V on both channels, frequency range 1 kHz to 13 MHz, Amplitude Function B/A and Phase reference A. In the B/A mode the instrument measures the relative amplitude Log(B/A) of the two input signals over a display range -100.0 dB to $+100$ dB and resolution 0.1 dB. The phase measurement range was $-180°$ to $+180°$ with 0.1° resolution. The GPM was set such that the

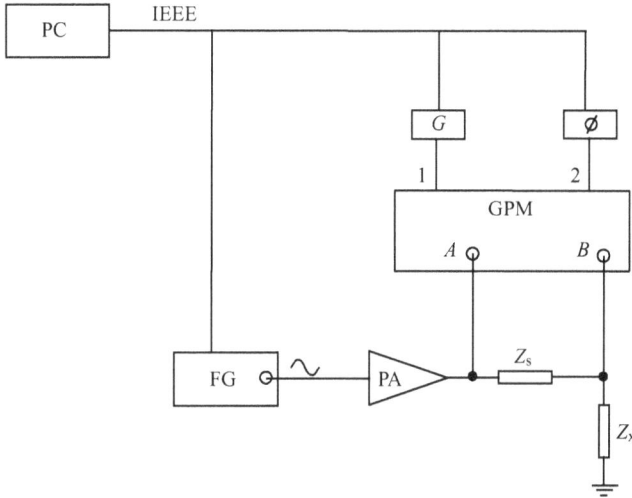

Figure 15.1   Measurement system. PC-computer with IEEE interface, FG, function generator; PA, power amplifier; GPM, gain phase meter. G and ϕ are DMM's measuring dc outputs from GPM proportional to V_B/V_A and phase, respectively.

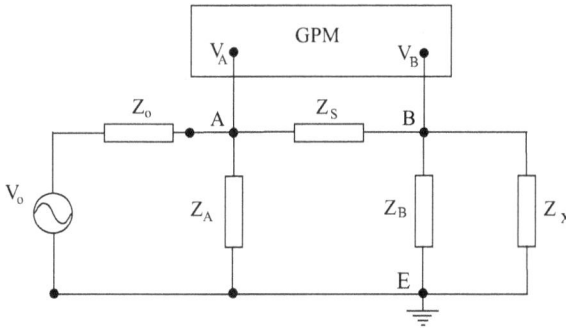

Figure 15.2   Equivalent impedance analysis circuit

analog output 1 corresponded to B/A with dc voltage 10mV/dB. Analog output 2 corresponded to the phase difference between *A* and *B* with dc voltage 10 mV/°.

The signal source was obtained from a function generator (FG) HP3325A controlled by a computer (PC) via an IEEE GPIB interface as shown in Figure 15.1. The output of FG was applied to a purpose built power amplifier (PA).

15.3.2   Equivalent circuit

The general equivalent circuit for the impedance analysis is shown in Figure 15.2. $\mathbf{V}_o$, $\mathbf{Z}_o$ are the voltage and impedance of the signal generator source. $\mathbf{Z}_A$ and $\mathbf{Z}_B$ are the input impedances of the GPM input channels. $\mathbf{Z}_s$ is a reference impedance

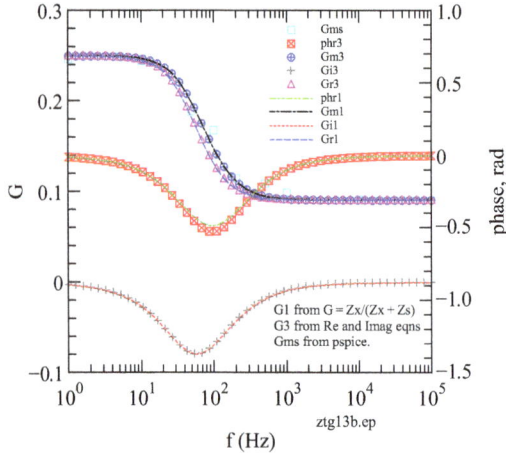

Figure 15.3   Frequency response of transfer functions. In this example the value of the components was: $R_s = 3\Omega$, $R_x = 1\Omega$, $L_s = 10\,mH$, $L_x = 1\,mH$.

and $\mathbf{Z}_x$ the unknown device impedance. In this figure GPM represents either a specific instrument or a dedicated IC such as the monolithic dual logarithmic amplifier AD8302.

The voltage at node A is

$$\frac{\mathbf{V}_A}{\mathbf{V}_o} = \frac{\mathbf{Z}_A//(\mathbf{Z}_s + \mathbf{Z}_B//\mathbf{Z}_x)}{\mathbf{Z}_o + \mathbf{Z}_A//(\mathbf{Z}_s + \mathbf{Z}_B//\mathbf{Z}_x)} \tag{15.1}$$

The voltages at nodes B and A determine the transfer coefficient $\mathbf{G} = \mathbf{V}_B/\mathbf{V}_A$ measured by the gain phase meter (Figure 15.3). This is given by

$$\mathbf{G} = \frac{\mathbf{V}_B}{\mathbf{V}_A} = \frac{\mathbf{Z}_B//\mathbf{Z}_x}{\mathbf{Z}_s + \mathbf{Z}_B//\mathbf{Z}_x} \tag{15.2}$$

$$\mathbf{G} = \frac{\mathbf{Z}_B\mathbf{Z}_x}{\mathbf{Z}_s\mathbf{Z}_B + \mathbf{Z}_s\mathbf{Z}_x + \mathbf{Z}_B\mathbf{Z}_x} \tag{15.3}$$

This equation may be simplified if we make $\mathbf{Z}_s \ll \mathbf{Z}_B$. Then,

$$\mathbf{G} = \frac{\mathbf{Z}_x}{\mathbf{Z}_s + \mathbf{Z}_x} \tag{15.4}$$

Expanding the complex impedances leads to

$$\mathbf{G} = \frac{R_x R_1 + X_x X_1 + j(X_x R_1 - X_1 R_x)}{R_1^2 + X_1^2} \tag{15.5}$$

where

$$R_1 = R_x + R_s, \quad X_1 = X_x + X_s \tag{15.6}$$

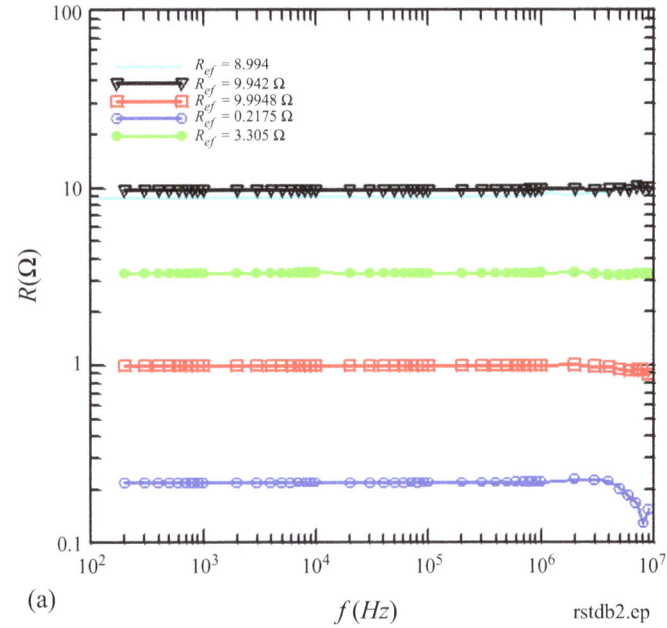

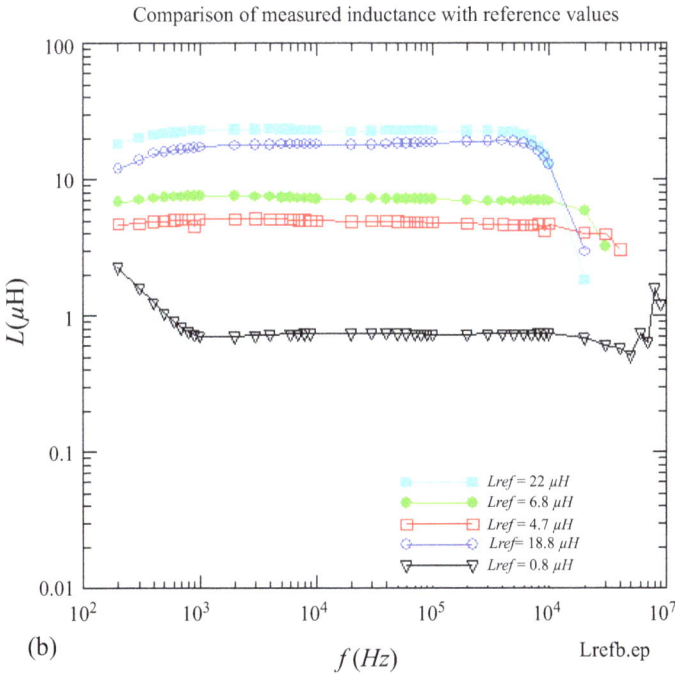

Figure 15.4   (a)Measured values of reference resistors using the GPM method and (b) measured values of reference inductors using the GPM method

The GPM measures the amplitude $|G| = V_A/V_B$ and phase angle ϕ between V_B and V_A. Thus the real G_r and imaginary terms G_i, amplitude $|G|$ and phase angle ϕ are

$$G_r = \frac{R_x R_1 + X_x X_1}{R_1^2 + X_1^2}, \quad G_i = \frac{X_x R_1 - R_x X_1}{R_1^2 + X_1^2}, \quad |G| = \sqrt{G_r^2 + G_i^2}, \quad \phi = tan^{-1}\frac{G_i}{G_r} \quad (15.7)$$

If we assume that the reactive terms X_x and X_s are the only frequency dependent components, that is, R_x and R_s are frequency independent then the frequency response of the amplitude and phase depend on functions of the form

$$|G| = F_1[X_x(f), X_s(f)], \quad \phi = F_2[X_x(f), X_s(f)] \quad (15.8)$$

For the skin effect case R_x also depends on frequency. Hence,

$$|G| = F_1[R_x(f), X_x(f), X_s(f)], \quad \phi = F_2[R_x(f), X_x(f), X_s(f)] \quad (15.9)$$

The impedance of the unknown is then given by rewriting (15.3) as follows:

$$\mathbf{Z}_x = \frac{\mathbf{Z}_s \mathbf{G}}{1 - \frac{\mathbf{G}(\mathbf{Z}_B + \mathbf{Z}_s)}{\mathbf{Z}_B}} \quad (15.10)$$

If we make $\mathbf{Z}_s \ll \mathbf{Z}_B$ or $\mathbf{Z}_x \ll \mathbf{Z}_B$ then

$$\mathbf{Z}_x = \frac{\mathbf{Z}_s \mathbf{G}}{1 - \mathbf{G}} \quad (15.11)$$

In the MATLAB® programme used to determine the impedance of the DUT the full (15.10) was used.

15.3.3 De-embedding

De-embedding was carried out as follows. With reference to Figure 15.2, the device to be measured is connected between terminals B and E, and the transfer function G and ϕ are measured. The device is replaced by a short circuit, and the measurements are repeated. With no device connected, the open-circuit impedance is $Z_{oc} = Z_B$, where Z_B is the input impedance of channel B of the GPM. The final unknown impedance is then given by

$$Z_x = \frac{Z_s G}{1 - G/G_{oc}} - \frac{Z_s G_{sc}}{1 - G_{sc}} \quad (15.12)$$

Thus, the unknown impedance Z_x is given in terms of the measured transfer function for the unknown device G, short circuit G_{sc}, open circuit G_{oc}, and the series impedance Z_s.

Results obtained using this GPM method are shown in Figures 15.4(a) and 15.4(b). The GPM accuracy using series impedance: $R_s = 3.3\,\Omega$, $L_s = 48\,$nH was better than 2% for the resistance measurements in a range 0.2–10 Ohms, frequency range 200 Hz to 1 MHz. For the inductance measurements, the accuracy was better than 10% over a range 1 μH to 20 μH, frequency range 1 kHz to about 0.5 MHz. Full details are given in Raven [120]. This includes skin effect measurements and using the technique to measure capacitors and diodes.

Appendix

A.1 Bessel's modified equation

For problems such as ac magnetic fields or current flow applied parallel to the z-axis of a conductor, we obtain partial differential equations of the form [76]

$$r^2 \frac{\partial^2 Y}{\partial r^2} + r \frac{\partial Y}{\partial r} - jm^2 Y r^2 = 0 \tag{A.1}$$

where $Y = f(r)$, r the radius of the cylinder and m is a constant. Now Bessel's equation with argument u and order v may be written as follows [78]:

$$u^2 \frac{\partial^2 Y}{\partial u^2} + u \frac{\partial Y}{\partial u} + (u^2 - v^2)Y = 0 \tag{A.2}$$

The general solution of this equation is

$$Y(r) = AJ_v(u) + BN_v(u) \tag{A.3}$$

where A and B are constants to be determined by the boundary conditions, $J_v(u)$ and $N_v(u)$ are Bessel functions of the first and second kind, respectively, with argument u and order v. If in (A.2), we substitute $u = rmj\sqrt{j}$ with $v = 0$, then (A.1) is obtained showing that equation (A.2) is a modified, zero order (v = 0) Bessel function. $J_v(u)$ is obtained from a series solution of the Bessel equation

$$J_v(u) = (u/2)^v [\frac{1}{v!} - \frac{(u/2)^2}{1!(v+1)!} + \frac{(u/2)^4}{2!(v+2)!} - \frac{(u/2)^6}{3!(v+3)!} + \cdots \tag{A.4}$$

$$= \frac{1}{\pi} \int_0^\pi \cos(n\theta - u\sin\theta) \, d\theta \tag{A.5}$$

$$N_v(u) = \frac{J_v(u)\cos v\pi - J_{-v}(u)}{\sin n\pi} \tag{A.6}$$

$N_v(u)$ is the Neumann function or Bessel function of the second kind. For non-integral v, this function satisfies Bessel's equation. However, if v is an integer, then the Neumann function is indeterminate and leads to non-physical solutions. It turns out that Neumann functions are not applicable for problems that involve the origin, such as axial ac currents or axial magnetic fields. Hence, for these problems we only require Bessel functions of the first kind, J_v. The solution to (A.1) is therefore

$$Y(r) = AJ_o(u) \tag{A.7}$$

Plot of Bessel functions order 0 and 1

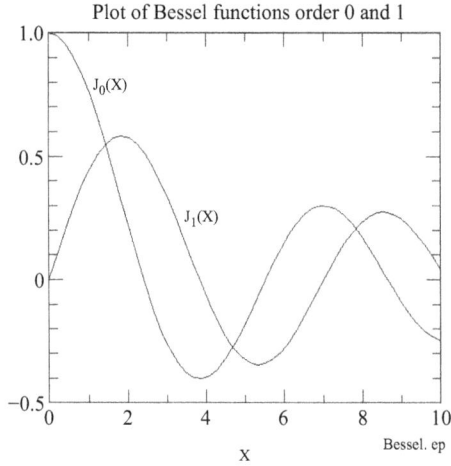

Figure A.1   Plot of zero order and first order Bessel functions, $J_o(x)$, $J_1(x)$, respectively, using MATLAB

The Bessel functions of the first kind, order zero and order one respectively are given by

$$J_o(u) = 1 - \frac{(u/2)^2}{(1!)^2} + \frac{(u/2)^4}{(2!)^2} - \frac{(u/2)^6}{(3!)^2} + \cdots = \sum_{m=0}^{\infty} (-1)^m \frac{(u/2)^{2m}}{(k!)^2} \qquad (A.8)$$

$$J_1(u) = \frac{u}{2} - \frac{(u/2)^3}{1!2!} + \frac{(u/2)^5}{2!3!} - \frac{(u/2)^7}{3!4!} + \cdots = \sum_{m=0}^{\infty} (-1)^m \frac{(u/2)^{2m+1}}{m!(m+1)!} \qquad (A.9)$$

These two orders are shown plotted in Figure A.1.

A.1.1   Kelvin functions

Equation (A.8) can be re-expressed in terms of Real and Imaginary complex components, referred to as Kelvin functions. The zero order Bessel function is then

$$J_o(u) = \Re J_o(u) + j\Im J_o(u) \qquad (A.10)$$

Substituting for $u = rmj\sqrt{j}$ in (8.5) gives

$$J_o(u) = 1 + j(mr/2)^2 - \frac{(mr/2)^4}{(2!)^2} - j\frac{(mr/2)^6}{(3!)^2} + \cdots \qquad (A.11)$$

Separating out the real and imaginary complex components gives the Kelvin functions

$$\Re J_o(u) = ber_o(mr) = 1 - \frac{(mr/2)^4}{(2!)^2} + \frac{(mr/2)^8}{(4!)^2} - \frac{(mr/2)^{12}}{(6!)^2} + \cdots \qquad (A.12)$$

$$\Im J_o(u) = bei_o(mr) = (mr/2)^2 - \frac{(mr/2)^6}{(3!)^2} + \frac{(mr/2)^{10}}{(5!)^2} \cdots \qquad (A.13)$$

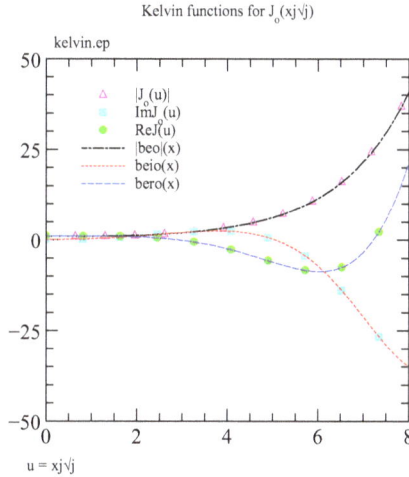

Figure A.2 Plots of Kelvin functions and real and imaginary Bessel functions from MATLAB

The Kelvin functions are shown plotted in Figure A.2.

Bessel functions of the first and second kinds with complex argument, that is, hyperbolic Bessel functions, are frequently expressed in terms of Kelvin functions I_n and K_n, respectively [6,76,79]. In this work we use the series solutions form for calculations involving MATLAB® and the Kelvin form for comparison with electrical equivalent circuits.

A.2 Properties of Bessel functions

The following formulae provide some useful relationships between Bessel functions which are found in many textbooks [73,78].

$$J_{-m}(u) = (-1)^m J_m(u), \quad m = integer \tag{A.14}$$

$$\frac{dJ_o(u)}{du} = -J_1(u) \tag{A.15}$$

$$\frac{d(uJ_1(u))}{du} = uJ_o(u) \tag{A.16}$$

$$\frac{dJ_n(u)}{du} = \frac{n}{u}J_n(u) - J_{n+1}(u) \tag{A.17}$$

$$\int uJ_o(u)du = uJ_1(u) \tag{A.18}$$

$$\int J_1(u)du = -J_o(u) \tag{A.19}$$

$$\int_0^{u_a} J_1(u)du = 1 - J_o(u_a) \tag{A.20}$$

$$\int u J_o^2(u) du = \frac{u^2}{2} [J_o^2(u) + J_1^2(u)] \tag{A.21}$$

In these cases $Z_n(u) = J_n(u)$ or $N_n(u)$, the Bessel function of the 2nd kind (Neumann function).

$$\frac{dZ_n(u)}{du} = \frac{n}{u} Z_n(u) - Z_{n+1}(u) \tag{A.22}$$

$$\int Z_1(u) du = -Z_o(u) \tag{A.23}$$

$$\int u Z_o(u) du = u Z_1(u) \tag{A.24}$$

$$\int u^2 Z_o^2(u) du = \frac{u^2}{2} [Z_o^2(u) + Z_o^2(u)] \tag{A.25}$$

$$\int u Z_n^2(u) du = \frac{u^2}{2} [Z_n^2(u) - Z_{n-1}(u) Z_{n+1}(u)] \tag{A.26}$$

A.3   Power integral

The integral in the power dissipation (6.3) and (6.7) requires solutions of the normal current density $J(r)$ and its complex conjugate $J^*(r)$.

For the good conductor, the complex conjugate current density is $J^*(r,t) = J_o e^{-j\omega t}$. The field distribution is obtained from Maxwell's equations, which yields the diffusion equation $\nabla^2 J^* = -jm^2 J^*$ where $m^2 = \omega\mu\sigma$. μ and σ are assumed constants. This may be expressed in cylindrical co-ordinates, assuming again that J^* does not vary with θ or z:

$$r^2 \frac{\partial^2 J^*}{\partial r^2} + r \frac{\partial J^*}{\partial r} + jm^2 J^* r^2 = 0 \tag{A.27}$$

Comparing this with Bessel's equation order zero

$$u_2^2 \frac{\partial^2 J^*}{\partial u_2^2} + u_2^2 \frac{\partial J^*}{\partial u_2^2} + u_2^2 J^* = 0 \tag{A.28}$$

where

$$u_2^2 = r^2 m^2 j, \quad u_2 = rmj^{1/2} \tag{A.29}$$

Previously, for the normal current density J we obtained $u = rmj^{3/2} = u_1$ say.

The solutions of Bessel's equation are expected to be orthogonal [6], provided appropriate boundary conditions are satisfied. In the case considered here, $r = a$, the radius of the cylinder is one boundary, and $r = 0$ is the other. Bessel's equation then becomes

$$r \frac{d^2 J_o(u_1)}{dr^2} + \frac{dJ_o(u_1)}{dr} + \left(\frac{u_1^2}{r}\right) J_o(u_1) = 0 \tag{A.30}$$

$$r \frac{d^2 J_o(u_2)}{dr^2} + \frac{dJ_o(u_2)}{dr} + \left(\frac{u_2^2}{r}\right) J_o(u_2) = 0 \tag{A.31}$$

where

$$u_1 = mrj^{3/2}, \quad u_2 = mrj^{1/2} \tag{A.32}$$

Note that here the general solution is $J(r) = CJ_o(u)$. Hence, $J(r)$ can be replaced with $J_o(u)$ because C is a constant. By using the differential product rule, the above two differential equations can be written as follows:

$$\frac{d}{dr}\left[r\frac{dJ_o(u_1)}{dr}\right] + \left(\frac{u_1^2}{r}\right)J_o(u_1) = 0 \tag{A.33}$$

$$\frac{d}{dr}\left[r\frac{dJ_o(u_2)}{dr}\right] + \left(\frac{u_2^2}{r}\right)J_o(u_2) = 0 \tag{A.34}$$

Multiply (A.31) by $J_o(u_2)$ and (A.33) by $J_o(u_1)$ gives

$$J_o(u_2)\frac{d}{dr}\left[r\frac{dJ_o(u_1)}{dr}\right] + \left(\frac{u_1^2}{r}\right)J_o(u_1)J_o(u_2) = 0 \tag{A.35}$$

$$J_o(u_1)\frac{d}{dr}\left[r\frac{dJ_o(u_2)}{dr}\right] + \left(\frac{u_2^2}{r}\right)J_o(u_2)J_o(u_1) = 0 \tag{A.36}$$

Subtract, noting that $\frac{u_2^2 - u_1^2}{r} = 2jm^2$

$$J_o(u_2)\frac{d}{dr}\left[r\frac{dJ_o(u_1)}{dr}\right] - J_o(u_1)\frac{d}{dr}\left[r\frac{dJ_o(u_2)}{dr}\right] = 2jm^2 rJ_o(u_1)J_o(u_2) = 0 \tag{A.37}$$

Integrating from $r = 0$ to $r = a$,

$$\int_o^a J_o(u_2)\frac{d}{dr}\left[r\frac{dJ_o(u_1)}{dr}\right]dr - \int_o^a J_o(u_1)\frac{d}{dr}\left[r\frac{dJ_o(u_2)}{dr}\right]dr$$

$$= 2jm^2\int_o^a rJ_o(u_1)J_o(u_2)\,dr \tag{A.38}$$

Integrating by parts ($\int uv'\,dx = uv - \int u'v\,dx$) gives for the left hand side

$$\int J_o(u_2)\frac{d}{dr}\left[r\frac{dJ_o(u_1)}{dr}\right]dr = J_o(u_2)\left[r\frac{dJ_o(u_1)}{dr}\right]$$
$$- \int\frac{J_o(u_2)}{dr}\left[r\frac{dJ_o(u_1)}{dr}\right]dr \tag{A.39}$$

$$\int J_o(u_1)\frac{d}{dr}\left[r\frac{dJ_o(u_2)}{dr}\right]dr = J_o(u_1)\left[r\frac{dJ_o(u_2)}{dr}\right]$$
$$- \int\frac{J_o(u_1)}{dr}\left[r\frac{dJ_o(u_2)}{dr}\right]dr \tag{A.40}$$

Subtracting, the right hand sides cancel. Hence, inserting limits gives the power integral

$$2jm^2 \int_o^a rJ_o(u_1)J_o(u_2) \, dr = \left[rJ_o(u_2)\frac{dJ_o(u_1)}{dr} \right]_o^a$$
$$- \left[rJ_o(u_1)\frac{dJ_o(u_2)}{dr} \right]_o^a \tag{A.41}$$

Now

$$\frac{dJ_o(u_1)}{dr} = \frac{dJ_o(u_1)}{du_1}\frac{du_1}{dr} = mj^{3/2}J_o'(u_1) = -mj^{3/2}J_1(u_1) \tag{A.42}$$

$$\frac{dJ_o(u_2)}{dr} = \frac{dJ_o(u_2)}{du_2}\frac{du_2}{dr} = mj^{1/2}J_o'(u_2) = -mj^{1/2}J_1(u_2) \tag{A.43}$$

$$2jm^2 \int_o^a rJ_o(u_1)J_o(u_2) \, dr = \left[mj^{3/2}rJ_o(u_2)J_o'(u_1) \right]_o^a$$
$$- \left[mj^{1/2}rJ_o(u_1)J_o'(u_2) \right]_o^a \tag{A.44}$$

This equation agrees essentially with Werner p. 207, for $v = 0$, $r = x$.

Bibliography

[1] Raven M. S. Experimental measurements of the skin effect and internal inductance at low frequencies, *Acta Technica*, vol. 60; 2015. pp. 51–69.

[2] Skin Effect — https://en.wikipedia.org/wiki/Skin_effect

[3] Maxwell J. C. *Treatise on Electricity and Magnetism* Vol. 2. Oxford University Press, Oxford; 1892.

[4] Penrose R. *The Road to Reality*. Vintage Books, London; 2004, p. 441.

[5] Einstein A. *Relativity-the Special and General Theory*. 14th Edn. London: Methuen; 1968, pp. 41, 49.

[6] Arfken G. *Mathematical Methods for Physicists*, New York: Academic Press; 1970.

[7] Raven M. S. Measuring low impedance and skin effect with a gain-phase meter, *Acta Technica*, vol. 59; 2014. pp. 303–320.

[8] Hampshire D. P. A derivation of Maxwell's equations using the Heaviside notation. Philosophical Transactions of the Royal Society A; 2018. https://doi.org/10.1098/rsta.2017.0447

[9] Mazur V. Principles of Lightning Physics, IOP Expanding Physics, Institute of Physics, UK; 2016, pp. 2054–7315. https://en.wikipedia.org/wiki/Lightning https://www.nssl.noaa.gov/education/svrwx101/lightning/

[10] Faria J. A. *Electromagnetic Foundations of Electrical Engineering*. Wiley-Interscience New York; 2008, p. 345.

[11] Audience Wave. https://en.wikipedia.org/wiki/Wave(audience).

[12] Landau L. D., Lifshitz, E. M., and Pitaeskii, L. P. *Electrodynamics of Continuous Media*, 2nd Edn. Butterworth Heinemann, London; 1996.

[13] Edwards J. and Saha T. K. Diffusion of Current into Conductors; 2001, pp. 401–406, AUPEC 01, Perth.

[14] BGR. Bundesanstalt für Geowissenschaften und Rohstoffe – Electromagnetics – Transient electromagnetics (TEM) 26 April 2019. https://www.bgr.bund.de/EN/Themen/GGGeophysik/Bodengeophysik/TransientenEM/teminhalten.htmlnn=1558406 Also see https://en.wikipedia.org/wiki/Transient-electromagnetics

[15] Crotti G., Giordano D., Roccato P., *et al.*, Pantograph-to-OHL Arc: Conducted Effects in DC Railway Supply System, *IEEE Transactions on Instrumentation and Measurement*, vol. 68, no. 10; 2019, pp. 3861–3870. https://doi.org/10.1109/TIM.2019.2902805

[16] Goossens M., Mittelbach F., and Samarin A. *The Latex Companion*. Addison-Wesley, Reading, MA; 1994, p. 183.

[17]   Christopoulos C. *The Transmission Line Modelling Method-TLM*, 1st Edn. IEEE Press, New York; 1995.

[18]   Maxwell, J. C. A Dynamical Theory of the Electromagnetic Field, *Philosophical Transactions of the Royal Society of London*, vol. 155; 1865, p. 49.

[19]   Kenyon I. R. *The Light Fantastic*, 2nd Edn. Oxford University Press, Oxford; 2011, p. 7.

[20]   Parton, J. E., Owen, S. J. T., and Raven, M. S. *Applied Electromagnetics*, 2nd Edn. Springer, Berlin; 1986.

[21]   Wang Z. L. General solutions of the Maxwell's equations for a mechano-driven media system (MEs-f-MDMS), *Journal of Physics Communication*, vol. 8; 2024, p. 115004. https://doi.org/10.1088/2399-6528/ad8d2f

[22]   Reitz J. R., Milford F. J., and Christy R. W. *Foundations of Electromagnetic Theory*, 3rd Edn. Addison-Wesley, Reading, MA; 1979 [see also Wikipedia – Gauge Fixing].

[23]   Crank J. *The Mathematics of Diffusion*. Clarendon Press, Oxford; 1975.

[24]   Maxwell, J. C. A Dynamical Theory of the Electromagnetic Field, *Philosophical Transactions of the Royal Society of London*, vol. 155; 1865, pp. 450–512.

[25]   Bureau International des Poids at Mesures (BIPM) SI Brochure, 9th Edn. 2019. https://www.bipm.org/en/publications/si-brochure/ [also see 'A concise summary of the International System of Units', SI-Brochure-9-concise-EN.pdf].

[26]   Rosa, E. B. Philosophical Magazine, (Ser 5) vol. 28; 1889, p. 315.

[27]   Parton J. E. Electrical Networks, pp. 1.13–1.15, 1974. Text book (unpublished) based on a course of about twenty lectures given to 2nd Year students, Dept. Electrical and Electronic Engineering, University of Nottingham, UK.

[28]   Poynting, J. H. On the transfer of energy in the electromagnetic field, *Philosophical Transactions of the Royal Society A*, vol. 175; 1884, p. 343–361.

[29]   Poynting J. H. The Pressure of Light. The Inquirer; 1903, pp. 195–196 (Google Scholar).

[30]   Loudon R. and Baxter C. Contributions of John Henry Poynting to the understanding of radiation pressure, *Philosophical Transactions of the Royal Society A*, 2011, p. 468. http://doi.org/10.1098/rspa.2011.0573

[31]   Dufresne J.-L. La détermination de la constante solaire par Claude Matthias Pouillet, *La Météorologie*, vol. 60; 2008, pp. 36–43.

[32]   Akpootu D. O. and Gana N. N. Evaluation of solar constant using locally fabricated aluminium cylinder, *Advances in Applied Science Research*, vol. 4, no. 5; 2013, pp. 401–408. Available online at www.pelagiaresearchlibrary.com

[33]   Kopp G. and Lean J. L. A new, lower value of total solar irradiance: Evidence and climate significance, *Geophysical Research Letters*, vol. 38, 2011, p. L01706, http://doi.org/10.1029/2010GL045777

[34]   Plonsey R. and Collin R. E. *Principles and Applications of Electromagnetic Fields*. McGraw-Hill, New York; 1961, pp. 310, 321, 326.

[35] Raven M. S. Skin effect in the time and frequency domain – comparison of power series and Bessel function solutions, *Journal of Physics Communication*, vol. 2; 2018, p. 035028. https://doi.org/10.1088/2399 -6528/aab4a8

[36] Raven M. S. Maxwell's vector potential method, transient currents and the skin effect, *Acta Technology*, vol. 58; 2013, pp. 337–350.

[37] Faria J. A. and Raven M. S. On the success of electromagnetic analytical approaches to full time-domain formulation of skin effect phenomena, *Progress In Electromagnetics Research PIER M*, vol. 31; 2013, pp. 29–43.

[38] Ramo S., Whinnery J. R., and Van Duzer T. *Fields and Waves in Communication Electronics*, 2nd Edn. Wiley, New York; 1984.

[39] Coufal O. Current density in two parallel cylindrical conductors and their inductance, *Electrical Engineering*, vol. 99; 2017, pp. 519–523.

[40] Faria J. A. *Comments on* 'Current density in two parallel cylindrical conductors and their inductance', *Electrical Engineering*, vol. 100, 2018, pp. 1535–1536. https://doi.org/10.1007/s00202-017-0633-0

[41] Chambers R. G. *Elelectrons in Metals and Semiconductors*. Chapman and Hall, 1990, p. 28, *Classical Electromagnetic Radiation*, Academic Press, New York, p. 104(1965).

[42] CODATA Recommended Values of the Fundamental Physical Constants, 2002, Reviews of Modern Physics, vol. 77; 2005, pp. 1–107 (Institute of Physics Diary, 2008).

[43] Plonsey R. and Collin R. E. *Principles and Applications of Electromagnetic Fields*, McGrawHill, New York; 1961, pp. 325, 328.

[44] Campbell D. S. and Hayes J. A. *Capacitive and Resistaive Electronic Components*. Gordon and Breach, Yverdon-Switzerland; 1994, p. 317.

[45] Terman F. E. *Radio Engineers' Handbook*. McGraw-Hill, New York; 1943, p. 48.

[46] Combes P. F. *Microwave Components, Devices and Active Circuits*. John Wiley & Sons; 1987, p. 28.

[47] Reitz J. R., Milford F.J., and Christy R.W., Foundations of EM Theory, 3rd Edn. Adison-Wesley, 1979, pp. 243–245.

[48] Bleaney B. I. and Bleaney B. *Electricity and Magnetism*. Oxford University Press, Oxford; 1965, p. 494.

[49] Kaye G. W. C. and Laby T. H. *Tables of Physical and Chemical Constants*. Longman, Harlow UK; 1989, p. 136.

[50] Williamson I. A. D., Nguyen T. N., Wang Z. Suppresion of the skin effect in radio frequency transmission lines via gridded conductor fibers, *Applied Physics Letters*, vol. 108, no. 8; 2016, p. 083502. https://doi.org/10.1063/1.4942649

[51] London F. and London H. *The electromagnetic equations of the supraconductor*, Proceedings of the Royal Society (London), vol. A149; 1935, pp. 71–88.

[52] Gorter C. J. and Casimir H. On supraconductivity I. *Physica*, vol. 1; 1934, pp. 306–320

[53]   Ginzburg V. L. and Landau L. D. Zh. Eksp. Teor. Fiz., vol 20; 1950, pp. 1064–1082. (Collected papers of L. D. Landau, ed D.ter Haar, Gordon and Breach 1967, No. 3)

[54]   Bardeen J., Cooper L. N., and Schrieffer J. R. Microscopic theory of superconductivity, *Physical Review*, vol. 108; 1957, pp. 162–164.

[55]   Bednorz J. G. and Muller K. A. Possible high Tc superconductivity in the Ba-La-Cu-O system. Z. Phys, B, vol. 64; 1986, p. 189.

[56]   Buzea C. and Robbie K. *Superconductor Science and Technology*, vol. 18; 2005, pp. R1–R8.

[57]   Poynting J. H., *Philosophical Transactions A*, vol. 175; 1884, p. 343.

[58]   Zhou S. *Electrodynamics Theory of Superconductors*. Peter Peregrinus Ltd., London; 1991, p. 45.

[59]   Duffin W. J. *Electricity and Magnetism*. McGraw-Hill, London; 1990, p. 355.

[60]   Corbin J. C. Skin effect, *Skin Effect in The Encyclopedia of Physics*, 3rd Edn. Ed Robert M. Besancon, Van Nostrand Reinhold, New York; 1990, p. 1119.

[61]   Starling S. G. and Woodall A. J. *Electricity and Magnetism for Degree Students*, 8th Edn. Longmans, London; 1953, p. 364.

[62]   Lord Rayleigh (John William Strutt), On the self-induction and resistance of straight conductors, Philosophical Magazine (Ser 5) vol. 21; 1886, p. 381.

[63]   Fleming J. A. Proceedings of the Physical Society of London, vol. 28; 1911, p. 103. Also see: A Note on the Experimental Measurement of the High-Frequency Resistance of Wires J A Fleming, Proceedings of the Physical Society of London, vol. 23; 1910, pp. 103–116. https://doi.org/10.1088/14 78-7814/23/1/311 Quantitative Measurements in Connection with Radio-Telegraphy, JIEE, XLIV, 349

[64]   MacDougal J. W. An experiment on skin effect, *American Journal of Physics*, vol. 44; 1976, p. 978.

[65]   Kennelly A. E. and Affel H. A. *Skin effect resistance measurements of conductors at radio frequencies up to 100,000 cycles per second*, Proceedings of the I.R.E., vol.58; 1916, pp. 523–574.

[66]   Faria J. A. and Raven M. S. On the success of electromagnetic analytical approaches to full time-domain formulation of skin effect phenomena, *Progress in Electromagnetics Research PIER M*, vol. 31; 2013, pp. 29–43.

[67]   Smith G. S. *A simple derivation for the skin effect in a round wire*, *European Journal of Physics*, vol. 35; 2014, pp. 1–13.

[68]   Riba J-R. Calculation of the ac to dc resistance ratio of conductive nonmagnetic straight conductors by applying FEM simulations, *European Journal of Physics*, vol. 36; 2015, pp. 1–10.

[69]   Starling S. G. and Woodall A. J. *Electricity and Magnetism*. Longmans and Co, London; 1956, p. 365.

[70]   Kraus J. D. *Electromagnetics*. McGraw-Hill, New York; 1992.

[71]   Silvester P. P. and Ferrari R. L. *Finite Elements for Electrical Engineers*. Cambridge University Press, Cambridge; 1983.

[72]   Chambers R. G., *Electrons in Metals and Semiconductors*. Chapman and Hall, London; 1990, p. 161.

[73]   Pozer D. M. *Microwave Engineering*. Adison-Wesley, New York; 1990.

[74]   Astbury N. F. *Electrical Applied Physics*. Chapman and Hall, London; 1956.

[75]   Assis A. K. T. and Hernandes J. A. *The Electric Force of a Current*. Aperion, Montreal; 2007, p. 66.

[76]   Werner Rosenheinrich Tables of some indefinite integrals of Bessel functions, Ernst Abbe Hochschule, Jena; 2015 (19.09.2015, First variant: 24.09.2003) University of Applied Sciences, Germany; 2003, pp. 207, 217.

[77]   Raven M. S. Skin effect in the time and frequency domain – comparison of power series and Bessel function solutions, *Journal of Physics Communications*, vol. 2, no. 3; 2018, p. 035028. https://doi.org/10.1088/2399-6528/aab4a8

[78]   Marion J. B. and Heald M. A. *Classical Electromagnetic Radiation*. Academic Press, London; 1965, p. 83.

[79]   Croxton C. A. *Introductory Eigenphysics*. John Wiley, London; 1974, p. 242.

[80]   Gormory F. Characteristics of high temperature superconductors by ac susceptibility measurements, *Superconductor Science and Technology* vol. 10; 1997, p. 523.

[81]   Couach M. and Khoder A. F. Magnetic Susceptibiity of Superconductors and other Spin Systems, Hein R. A. et al.,(Editors) Plenum, New York; 1991, p. 25.

[82]   Raven M. S. and Salim M. Design aspects of a differential magnetic susceptometer for high temperature superconductors, *Measurement Science and Technology* vol. 12; 2001, pp. 744–754.

[83]   Duffin W. J. *Electricity and Magnetism*. McGraw-Hill, London; 1990, p. 240, Problem 9.9.

[84]   Salim M. The AC Magnetic Susceptibility of High temperature Superconductors, PhD Thesis, The University of Nottingham; 2001.

[85]   Gueffaf A. Paraconductivity and Excess Hall Effect of YBCO Thin Films, PhD Thesis, The University of Nottingham; 2001.

[86]   Raven M. S. Thin Film CouplingCoupling2.doc; 2007 (Unpublished).

[87]   www.Leonardo-energy.org

[88]   Smythe W. R. *Static and Dynamic Electricity*, 2nd Edn. McGraw-Hill, New York; 1950.

[89]   King D. W. Practical continuous functions for the internal impedance of solid cylindrical conductors; 2012. http://www.g3ynh.info/

[90]   Payne A. https://www.researchgate.net/publication/351312897-THE-AC-RESISTANCE-AND-INDUCTANCE-OF-RAILS; 2021.

[91]   Raven M. S. Experimental measurements of the skin effect and internal inductance at low frequencies. *Acta Technica*, vol. 60; 2015, pp. 51–69.

[92]   Pozar D. M. *Microwave Engineering*. Addison-Wesley, New York; 1990.

[93]   Kanthal Handbook, p. 15, Sandvik, www.Kanthal.com

[94]   Raven M. S. Axial Impedance of Cylindrical Conductors – V/I approach.

[95]   Raven M. S. Skin effect in the time and frequency domain – comparison of power series and Bessel function solutions, *Journal of Physics Communications*, vol. 2; 2018, p. 035028. https://doi.org/10.1088/2399-6528/aab4a8

[96]   The copper wire used for the electrical measurements was standard electrical wire used in house wiring. The main grade of copper used for electrical applications is electrolytic-tough pitch (ETP) copper (CW004A or ASTM designation C11040). This copper is at least 99.90% pure and has an electrical conductivity of at least 101% IACS. UK std BS 6722:1986. https://web.archive.org/web/20130523163147/, http://www.ndt-ed.org/GeneralResources/IACS/IACS.htm, https://www.copper.org/, https://en.wikipedia.org/wiki/Copper-conductor

[97]   Aluminium: Kaye G. N. C. and Laby T. L. (Editors). *Tables of Physical and Chemical Constants*, 15th Edn. Longman Scientific and Technical; 1989, p. 178. Lide D. R. (Editor), *CRC Handbook of Chemistry and Physics*, 82nd Edn. CRC Press; 2001–2002, pp. 12–45. Tennent R. M. (Editor). *Science Data Book*. Open University, Oliver and Boyd; 1974, p. 60. Wikipedia https://en.wikipedia.org/wiki/Aluminium

[98]   Raven M. S. Skin effect in the time and frequency domain – comparison of power series and Bessel function solutions, *Journal of Physics Communications*, vol. 2; 2018, p. 035028 https://doi.org/10.1088/2399-6528/aab4a8

[99]   Dampier W. C. *A History of Science*. Cambridge University Press, Cambridge; 1966, p. 42.

[100]   Aurubis, Cu-DPH Material datasheet, EN-2024-06, aurubis.com/stolberg

[101]   Substations - Wikipedia.htm

[102]   https://www.totalconnections2009.co.uk/article/copper-tube-busbars/

[103]   https://en.wikipedia.org/wiki/Hayes-substation-fire

[104]   Raven M. S. *Measuring low impedance and skin effect with a gain-phase meter*, *Acta Technica*, vol. 9; 2014, pp. 303–320.

[105]   Raven M. S. Impedance and Skin Effect Measurements for a Large Regular Planar Copper Wire Meander, *Acta Technica*, vol. 61; 2016, pp. 91–105.

[106]   Upadhye A. and El-Sharkawi M. Cable-Properties: Computation of Cable Properties and ATP Simulations.

[107]   Crank J. *The Mathematics of Diffusion*, 2nd Edn. Clarendon Press, Oxford; 1975.

[108]   Hayt W. H. *Engineering Electromagnetics*, 4th Edn. McGraw-Hill, New York; 1981.

[109]   WWW.ACAsolutions.com or 1.800.866.7385.

[110]   Kraus J. D. *Electromagnetics*, 4th Edn. McGraw Hill, New York; 1991.

[111]   Kvitkovic J., Pamidi S., and Voccio J. Shielding AC magnetic fields using commercial YBa2Cu3O7-coated conductor tapes, *Superconductor Science and Technology*, vol. 22; 2009.

[112]   Zheng Z. and Zhang R. Metal detecting sensor based on linear hall effect elements, *Applied Mechanics and Materials*, vols. 530–531; 2014, pp. 83–90.

[113]   Spegel-Lexne D., Gómez S., Argillander J., Pawlowski M., and Xavier G. B., Experimental demonstration of the equivalence of entropic uncertainty with wave-particle duality, *Science Advances*, vol. 10, no. 49; 2024. https://doi.org/10.1126/sciadv.adr2007

[114]   Barnett S. M. The quantum optics of media, *Philosophical Transactions of the Royal Society A*; 2024, p. 382. https://doi.org/10.1098/rsta.2023.0339

[115]   Faria J. A. Skin effect in inhomogeneous Euler-Cauchy tubular conductors, *Progress in Electromagnetics Research M*, vol. 18; 2011, pp. 89–101.

[116]   Raymond Calvert R. The Transformer Ratio-Arm Bridge. Wayne Kerr Monograph No. 1.

[117]   Rogal B. *Recent advances in three-terminal bridge techniques*, Proceedings of the Institution of Electronics vol. 4, no. 2; The Institution of Electronics, London; 1961, pp. 8–14.

[118]   Impedance Measurement Handbook - Keysight www.keysight.com

[119]   Prabhakaran S. and Sullivan C. R.: *Impedance-Analyzer Measurement of High-Frequency Power Passives: Techniques for High Power and Low Impedance*, IEEE Industry Applications Society Annual Meeting; 2002. p. 1360–1367.

[120]   Raven M. S. Measuring low impedance and skin effect with a gain-phase meter, *Acta Technica*, vol. 59; 2014. pp. 303–320.

[121]   Li N., Xu H., Wang W., Zhou Z., Qiao G. and D-U Li D. A high-speed bioelectrical impedance spectroscopy system based on the digital auto-balancing bridge method, *Measurement Science and Technology*, vol. 24; 2013, pp. 1–12.

[122]   Tomasz Piasecki. Fast impedance measurements at very low frequencies using curve fitting algorithms, *Measurement Science and Technology*, vol. 26; 2015, pp. 1–9.

[123]   Angrisani L., Baccigalupi A. and Pietrosanto A. A digital signal-processing instrument for impedance measurement, *IEEE Transactions on Instrumentation and Measurement*, vol. 45; 1996, pp. 930–934.

[124]   Gaydecki P., Miller G., Zaid M. and Fernandes B. Detection of magnetic fields highly attenuated by the skin effect through a ferrous steel boundary using a super narrow-band digital filter, *IEEE Transactions on Instrumentation and Measurement*, vol. 57; 2008, pp. 1171–1176.

[125]   Sun S., Xu L., Cao Z., Zhou H. and Yang W. A high-speed electrical impedance measurement circuit based on information-filtering demodulation, *Measurement Science and Technology*, vol. 25; 2014, pp. 1–10.

[126]   Dumbrava V. and Svilainis L. The automated complex impedance measurement system, *Electronics and Electrical Engineering*, vol. 4, no. 76; 2007, pp. 59–62.

[127]   Janeiro F.M. and Ramos P.M. Impedance measurements using genetic algorithms and multiharmonic signals, *IEEE Transactions on Instrumentation and Measurement*, vol. 58; 2009, p. 383.

[128]   Sanchez B., Vandersteen G., Bragos R. and Schoukens J. Basics of broadband impedance spectroscopy measurements using periodic excitations, *Measurement Science and Technology*, vol. 23; 2012.

[129]   Lewis Jr G. K., Lewis Sr G. K. and Olbricht W. Cost-effective broad-band electrical impedance spectroscopy measurement circuit and signal analysis for piezo-materials and ultrasound transducers, *Measurement Science and Technology*, vol. 19; 2008, pp. 102–105.

[130]   Muciek A. and Cabiati F. Analysis of a three-voltmeter measurement method designed for low-frequency impedance comparisons, *Metrology and Measurement Systems*, vol. 13; 2006, pp. 19–33.

[131]   Callegaro L., Galzerano G. and Svelto C. Precision impedance measurements by the three-voltage method with a novel high-stability multiphase DDS generator, *IEEE Transactions on Instrumentation and Measurement*, vol. 52; 2003, pp. 1195–1199.

[132]   Abraham M. and Rajasekaran K. Implementation of bioimpedance instrument Kit in ARM7, *International Journal of Advanced Research in Computer Science and Software Engineering*, vol. 3; 2013, pp. 1271–1273.

[133]   Cowles J. and Gilbert B. Accurate gain/phase measurement at radio frequencies up to 2.5 GHz, *Analog Dialogue*, vol. 35; 2001, pp. 5–8.

[134]   Copper Development Association High Conductivity Copper for Electrical Engineering, copperalliance.org.uk/docs; 1998.

[135]   Ducluzaux A. Extra losses caused in high current conductors by skin and proximity effects, *Cahier Technique Schneider Electric*, no. 83; 1983. http://www.schneider-electric.com

Index